Arthur Parnell

The Action of Lightning and the Means of Defending Life and Property from Its Effects

Arthur Parnell

The Action of Lightning and the Means of Defending Life and Property from Its Effects

ISBN/EAN: 9783337094638

Printed in Europe, USA, Canada, Australia, Japan

Cover: Foto ©berggeist007 / pixelio.de

More available books at **www.hansebooks.com**

THE
ACTION OF LIGHTNING

AND

THE MEANS OF DEFENDING LIFE AND PROPERTY FROM ITS EFFECTS

BY

ARTHUR PARNELL
MAJOR IN THE CORPS OF ROYAL ENGINEERS

LONDON
CROSBY LOCKWOOD AND CO.
7, STATIONERS' HALL COURT, LUDGATE HILL
1882

PREFACE.

The object of this work is to present in some detail the probable nature of the action of lightning, and to suggest means which shall tend to give better and more economical defence to life and property against its effects than is afforded by the methods now in use.

The work is divided into three Parts.

Part I. consists of recorded facts and opinions. The latter have been compiled partly as supplying evidence of the ideas generally entertained on the subject, and partly for the purpose of fortifying the views now advanced.

Part II. is a superstructure of theory erected on the foundation of fact and electrical law comprised in the foregoing Part.

Part III. is the practical outcome of this theory, and describes the various defensive measures advocated.

DEVONPORT,
April 30*th*, 1881.

TABLE OF CONTENTS.

LIST OF AUTHORITIES.

PART I.—FACTS AND OPINIONS.

CHAPTER I.—NOTES ON LIGHTNING.

	PAGE
SECTION A.—ELECTRICAL NOTES	1
(a) Electricity	1
(b) Electrical Measurement	2
(c) Potential	3
(d) Insulation	4
(e) Induction	4
(f) Condensation	5
(g) Electric Sparks	6
SECTION B.—THE CONDUCTIVITY OF MATERIALS	6
SECTION C.—THE ELECTRICITY AND MAGNETISM OF THE EARTH	10
(a) Atmospheric Electricity	10
(b) Rain and Hail	12
(c) Clouds	13
(d) Terrestrial Electricity	14
(e) Earth Currents	16
(f) The Earth's Magnetism	17
(g) Auroræ	20
(h) St. Elmo's Fires	21
(i) Waterspouts	22
(k) Earthquakes	23
(l) Volcanic Eruptions	25

	PAGE
SECTION D.—LIGHTNING DISCHARGES	26
(a) The Nature of Lightning	26
(b) The Action of Lightning on Materials	30
(c) Return Strokes	31
(d) The Effect of Lightning on Persons	32
(e) The Effect of Lightning on Telegraphs	33
SECTION E.—THE INFLUENCE OF METALS ON LIGHTNING	35
SECTION F.—PRESERVATIVES FROM LIGHTNING	36
SECTION G.—STATISTICAL AND GEOGRAPHICAL NOTES	39

CHAPTER II.—NOTES ON LIGHTNING ENGINEERING.

SECTION A.—HISTORICAL NOTES	47
SECTION B.—DETAILS OF LIGHTNING RODS	52
SECTION C.—POINTS OF RODS	59
SECTION D.—EARTH CONNECTIONS OF RODS	65
SECTION E.—THE APPLICATION OF RODS	70
SECTION F.—THE INSPECTION OF RODS	75
SECTION G.—THE PROTECTIVE POWERS OF RODS	77

CHAPTER III.— SOME INCIDENTS OF LIGHTNING ACTION.

CHAPTER IV.—SOME INSTANCES OF EXISTING LIGHTNING RODS.

PART II.—THE THEORY OF THE ACTION OF LIGHTNING.

CHAPTER V.—ELECTRICAL DEFINITIONS AND DATA.

SECTION A.—ELECTRICAL DEFINITIONS	145
(a) Fundamental Terms	145
(b) The Influence of Bodies	147
(c) The Nature of Condensers	147

	PAGE
SECTION B.—ELECTRICAL DATA	148
(a) Electrical Formulæ	148
(b) The Three Elements of Electricity	149
(c) Collectors and Insulators	150
(d) Electrical Explosions	150
(e) Electrical Return Strokes	151
(f) Electrical Leaks	151
(g) Illustrations of Electrical Action	152

CHAPTER VI.—THE CONSTITUTION OF THE TERRESTRIAL CONDENSER.

SECTION A.—THE FUNCTION OF THE EARTH IN THE TERRESTRIAL CONDENSER	154
(a) The Relation between the Earth and the Clouds	154
(b) The Earth's Electricity	156
SECTION B.—THE THEORY OF DESCENDING LIGHTNING	160
(a) Facts Regarding Descending Lightning	160
(b) Descending Lightning from the Aspect of Electrical Law	165
SECTION C.—THE OUTLINE OF THE TERRESTRIAL PLATE	167
SECTION D.—THE INFLUENCES OF THE MATERIALS COMPOSING THE TERRESTRIAL CONDENSER	170
(a) Table of Influences of Various Substances	171
(b) Remarks on the Table	174
SECTION E.—THE DISCHARGE OF THE TERRESTRIAL CONDENSER	174
(a) The Various Forms of Terrestrial Electrical Discharge	174
(b) The Rationale of Thunderbolts	176

CHAPTER VII.—THE ACTION OF THUNDERBOLTS.

SECTION A.—THE ELECTRICAL CONDITIONS OF THE EARTH'S SURFACE	179
(a) The Accumulation of Electricity on the Earth's Surface	179
(b) Surfaces of Water	180
(c) Moist Earth	182
(d) Rocky and Dry Surfaces	182
(e) Paved Surfaces	183
(f) Surfaces Formed by Railway Metals	185
(g) The Shape and Geological Formation of the Ground	185
(h) Analysis of Incidents in regard to Conditions of Surface	186

	PAGE
SECTION B.—DETAILS OF THUNDERBOLT ACTION	187
(a) Classification of Objects on the Earth's Surface	187
(b) Electrical Connection	189
(c) Explosive Action	190
(d) Local Plates	194
(e) Persons in the Open Air	196
(f) Local Dielectrics	198
(g) Accidental Dielectrics Formed by Local Plates	201
(h) The Protection Afforded by the Interiors of Buildings	203
(i) The Dangers to which Interiors are Liable	204
(k) The Special Danger from Chimneys	206
(l) Simultaneous Strokes of Lightning	209
SECTION C.—ANALYSIS OF THUNDERBOLT INCIDENTS	209
(1) Buildings	210
(2) Ships	212
(3) Metals	213
(4) Chimneys	215
(5) Trees	215
(6) Flagstaffs, Masts, &c.	216
(7) Telegraphs	216
(8) Persons	216
(9) Animals	218
(10) Simultaneous Strokes	218
(11) Repeated Strokes	219
(12) Accurately defined Strokes	219
(13) Horizontally directed portions of Strokes	219
(14) Acts of Mechanical Force (except to Lightning Rods) exclusive of Rending of Masonry	219
(15) Objects which, when Struck, probably formed Local Plates	219
(16) Objects which, when Struck, probably formed Local Plates, associated with Local Dielectrics	220
(17) Objects which, when Struck, probably formed Local Plates, accidentally constituting Local Dielectrics	220
(18) Objects which, when Struck, probably formed Local Dielectrics	220
SECTION D.—ATMOSPHERIC DIELECTRICAL CONDITIONS	220
(a) The Influence of Rainfall	221
(b) The Temperature of the Air	222
(c) Atmospheric Electricity	223

		PAGE
SECTION E.—CLOUDS AND CLOUD EXPLOSIONS	. . .	223
(a) The Electrical Conditions of the Clouds .	. .	223
(b) The Electricity Due to the Conversion of the Clouds into Rain .	. . . .	224
(c) Cloud Explosions .	. . . . . .	225
(d) Thunderstorms	. . . .	226
SECTION F.—TERRESTRIAL RETURN STROKES	. . .	227
(a) Nature of Terrestrial Return Strokes	. . .	227
(b) Return Strokes Induced by Cloud Explosions .	. .	228
(c) Return Strokes Induced by Thunderbolts	. .	229
(d) The Effect of Return Strokes on Telegraphs .	. .	230
SECTION G.—TERRESTRIAL LEAKS	. . . . .	231
(a) Atmospheric Porous Leaks	. . . . .	231
(b) Terrestrial Angular Leaks	. . . . .	232
(c) The Value of Metal Points in Relation to the Earth	.	233
(d) Terrestrial Valves afforded by Features of Civilisation	.	235
(e) The Angular Leakage of Metals	. . . .	236

CHAPTER VIII.—THE PRESENT SYSTEM OF LIGHTNING RODS.

SECTION A.—OBSERVATIONS ON THE HISTORY OF LIGHTNING RODS .	. . . . . .	238
(a) The Invention of Lightning Rods	. . . .	238
(b) Theory of the Functions of Rods	. . . .	238
(c) The Opposition to the Use of Rods .	. . .	242
(d) Diverse Systems of Application	. . . .	244
SECTION B.—COMMENTS ON THE THEORY AND PRACTICE OF LIGHTNING ROD DEFENCE	. . .	245
(a) The Exposure of Elevated Metal	. . . .	246
(b) The Costliness of Rods	. . . . .	249
(c) The Sources of Failure to which Rods are Liable	.	254
(d) The Tendency of Rods to Disfigure Buildings .	.	257
SECTION C.—ANALYSIS OF LIGHTNING INCIDENTS CONNECTED WITH RODS	. . . . . .	258
(1) Rods Struck .	. . . .	258
(2) Constructions Struck, but not the Rods .	. .	259
(3) Constructions, without Rods, Damaged, Close to Other Constructions which had Rods, but Received no Damage	. . . .	259

(4) Buildings Struck before being Supplied with Rods, but not Subsequently	PAGE 259
(5) Causes to which the Failures of Rods have been Attributed by Authorities	259
(6) Rods which probably Acted as Local Plates . . .	260
(7) Rods which probably Acted as Accidental Dielectrics	260
(8) Rods which probably Acted as Local Dielectrics .	260
(9) Ends of Rods Fused	260
(10) Mechanical Injuries to Rods	260
(11) Deviations of the Explosions from the Rods . .	261
(12) Notes	261
SECTION D.—SUMMARY OF REMARKS ON LIGHTNING RODS .	264

PART III.—PRACTICAL MEASURES ADVOCATED FOR THE DEFENCE OF LIFE AND PROPERTY FROM THE EFFECTS OF LIGHTNING.

CHAPTER IX.—PRACTICAL MEASURES ADVOCATED.

SECTION A.—THE DEFENCE OF LARGE AREAS . . .	267
(a) The Defence of Countries and Districts . . .	267
(b) The Defence of Towns	269
SECTION B.—THE DEFENCE OF CONSTRUCTIONS . . .	270
(a) The Removal of Metal and Explosive Conditions .	271
(b) The Reduction of the Explosiveness of the Ground .	276
(c) The Conversion of Chimney Grates into Electric Taps	278
(d) The Application of Electric Taps to the Ground Surrounding the Building	280
(e) Summary of Proposals for the Defence of Buildings	285
(f) The Defence of Coal Mines	287
(g) The Defence of Ships	288
SECTION C.—THE DEFENCE OF INDIVIDUALS . . .	289
(a) Rules for the Guidance of Individuals . . .	289
(b) Agricultural Labourers	291

LIST OF AUTHORITIES.

Most of the notes and incidents in Chapters I., II., and III. have been extracted from the following authorities, which are respectively denoted by the accompanying abbreviations:—

(1.) "Encyclopædia Britannica." Edition 1857 . . *Enc. Br.*
(2.) "The Life of Benjamin Franklin, written by Himself." Edited by John Bigelow. London: Lippincott. 1879. *Frank.*
(3.) "Aide-Mémoire to the Military Sciences, framed from Contributions of Officers of the different Services, and edited by a Committee of the Corps of Royal Engineers." London: John Weale. 1853. Article on "Geognosy," by Colonel J. E. Portlock, R.E., F.R.S. . . *Portl. A. M.* Article on "Electricity," by Colonel R. J. Nelson, R.E.
Nels. A. M.
(4.) François Arago's "Meteorological Essays," translated by Colonel Sabine, R.A., F.R.S. London: Longmans. 1855. *Ar.*
(5.) "Papers on Subjects Connected with the Duties of the Corps of Royal Engineers." Vol. XVIII. Article by Lieut. T. Fraser, R.E., "How Earthquakes can be Observed and Registered." . . . . *Fras. R. E. P.*
(6.) "Aide-Mémoire for the use of Officers of Royal Engineers." Compiled by Colonel A. C. Cooke, C.B., R.E. Vol. I. London. 1879. . . . . . . *R. E. A.*
(7.) "Library of Useful Knowledge. Natural Philosophy." Vol. II. "Electricity," by Dr. Roget. London: Baldwin . . . . *Rog.*
(8.) "Elementary Treatise on Physics, Experimental and Applied." Translated and edited from "Ganot's Eléments de Physique," by E. Atkinson, Ph.D., F.C.S., Professor of Experimental Science, R.M.C., Sandhurst. Third Edition. London: Longmans. 1868. . . . . . *Gan.*
(9.) "The Medical Remembrancer, or Book of Emergencies," by Edward B. L. Shaw. Fourth Edition. London: Churchill. 1856. . . . . *Shaw.*
(10.) Journal of the Society of Telegraph Engineers for 1872.
S. T. E.
(11.) Herschel's Meteorology. A. and C. Black. 1862. . *Hersch.*
(12.) Kaemtz's Meteorology. Translated by C. V. Walker. 1845. . . . . *Kaem.*
(13.) "Electricity and Magnetism," by Fleeming Jenkin, F.R.S.S., L. and E., M.I.C.E., Professor of Engineering

in the University of Edinburgh. London: Longmans. 1873. . . . . *Jen.*

(14.) Culley's "Lectures on Construction and Maintenance of Telegraph Lines," delivered at the S.M.E., Chatham. February, 1869. . . . . *Cull.*

(15.) "The Atmosphere," by Camille Flammarion. Translated from the French, and edited by James Glaisher, F.R.S. London. 1873. . . . . *Flam.*

(16.) "On the Nature of Thunderstorms, and on the means of Protecting Buildings and Shipping against the Destructive Effects of Lightning," by W. Snow Harris, F.R.S. London: Parker. 1843. . . . . *Harr.*

(17.) "Lessons in Electricity," delivered at the Royal Institution in 1875-6, by John Tyndall, Professor of Natural Philosophy, Royal Institution of Great Britain. London: Longmans. 1876. . . . . *Tynd.*

(18.) "On the Protection of Buildings from Lightning," by R. J. Mann, M.D. A paper read at a meeting of the Society of Arts, 28th April, 1875, and published in their Journal. Letter from Dr. Mann to the *Times*, 23rd November, 1877. "Further Remarks concerning the Lightning Rod," by Dr. Mann. A paper read at a meeting of the Society of Arts, 15th March, 1878, and published in their Journal. . . . . *Mann.*

(19.) Chambers's "English Dictionary." 1877. . . *Chamb.*

(20.) "On Lightning and Lightning Conductors," by W. H. Preece, M.I.C.E. A paper read at a meeting of the Society of Telegraph Engineers, on the 27th November, 1872, and published in their Journal. Also other papers by the same author. . . . . *Preece.*

(21.) Inaugural address to the Society of Telegraph Engineers, on the 15th January, 1874, by Sir William Thomson, F.R.S., LL.D. (also other quotations from the same author). . . . . *Thoms.*

(22.) Registrar-General's Report for England and Wales 1877-1878. . . . . *Reg. Engl.*

(23.) Registrar-General's Report for Ireland, 1877-1878. *Reg. Irel.*

(24.) War Office Instructions as to the application of lightning conductors for the protection of powder magazines and other buildings. } { 24th July, 1829, 18th March, 1846, 25th May, 1858, 6th April, 1875. } *W. O.*

[NOTE.—The appendixes A and B to W.O. Instructions of 25th of May, 1858, are by Sir W. Snow Harris.]

(25.) "Rough Notes on Electricity." Chatham. 1873 . *Chath.*
(26.) "Lightning Conductors, their History, Nature, and Mode of Application," by Richard Anderson, F.C.S., F.G.S., Member of the Society of Telegraph Engineers. London: Spon. 1879 . . . . *And.*
(27.) "Encyclopædia of Experimental Philosophy." "Electricity," by the Rev. Francis Lunn, A.M., F.R.S. . *Lunn.*
(28.) Clark and Sabine's Electrical Tables and Formulæ . *Clark.*
(29.) Deschanel's "Natural Philosophy," edited by Professor Everett, F.R.S. 1872 . . . . . . *Desch.*
(30.) "How to Build a House," a translation by B. Bucknall, Architect, of Viollet-le-Duc's "Histoire d'une Maison." London: Sampson Low & Co. 1874 . . . *Violl.*
(31.) Buchan's "Handy Book of Meteorology." Second Edition. Edinburgh: Blackwood. 1868 . . . . *Buch.*
(32.) "Scrambles amongst the Alps in the Years 1860–69," by Edward Whymper. London: Murray. 1871. . *Whymp.*
(33.) Report of Major Majendie, R.A., H.M. Inspector of Explosives, to the Home Secretary, dated 17th September, 1878. . . . . *Rep. Expl.*
(34.) *Royal Engineers' Journal*, 1876. No. lxiii. . . *R.E.J.*
(35.) The *Wellington Weekly Gazette* . . . . . *W.W.G.*
(36.) The *Telegraphic Journal and Electrical Review* . . *Tel.*
(37.) Symons' *Meteorological Magazine* . . . . *S.M.M.*
(38.) The *Times* newspaper . . . . . . *Times.*
(39.) The *Illustrated London News* . . . . . *I.L.N.*
(40.) The *Standard* newspaper . . . . . . *Stand.*
(41.) The *Graphic* newspaper . . . . . . *Graph.*
(42.) The *Western Morning News* . . . . . *W.M.N.*
(43.) The *Western Weekly News* . . . . . *W.W.N.*
(44.) Kentish newspaper . . . . *K.P.*
(45.) Mr. Von Fischer Treuenfeld, M.S.T.E. . . . *Treu.*
(46.) M. Francisque Michel . . . . . *Franc. Mich.*
(47.) Mr. G. J. Symons, F.R.S. . . . . . *Sym.*
(48.) Mr. Latimer Clark, F.R.S. . . . . *Lat. Clark.*
(49.) Mr. James Graves, M.S.T.E. . . . . . *Grav.*
(50.) M. E. Nouel . . . . *Nouel.*
(51.) Captain D. Galton, F.R.S., C.B. . . . . *Galt.*
(52.) Professor Abel, F.R.S. . . . . . . *Abel.*
(53) Mr. W. E. Ayrton, M.S.T.E. . . . . . *Ayrt.*
(54.) "A Physical Treatise on Electricity and Magnetism," by J. E. H. Gordon, B.A. Camb., Assistant Secretary of the British Association. London: Sampson Low & Co. 1880 . . . . *Gord.*

⋮

ON LIGHTNING.

Part I.

FACTS AND OPINIONS.

CHAPTER I.—NOTES ON LIGHTNING.

(A.) Electrical Notes.

(a) *Electricity.*

(1.) ELECTRICITY is one form of energy, and therefore necessarily force, and not matter. (*Preece*, 337.)

(2.) "Electricity is a powerful physical agent which manifests itself mainly by attractions and repulsions, but also by luminous and heating effects, by violent commotions, and many other phenomena." (*Gan.* 584.)

(3.) "It may be impossible that we shall ever arrive at a perfect knowledge of the subtle operations from whence the phenomena of electricity result. Therefore any theoretical view of them, as of many other questions in physical science, is but a sort of intellectual contrivance for representing to the mind the order and connection subsisting between observed phenomena." (*Harr.* 66.)

(3*a*.) "We have as yet no conception of electricity apart from the electrified body; we have no experience of its independent existence." (*Gord.* i. 1.)

(4.) The science of electricity "has two great divisions; the one called 'Frictional electricity,' the other 'Voltaic electricity.'" (*Tynd.* 19.)

I. A b 5—8.

(5.) Electricity is derived from the following sources, viz. :—

1. Friction.
2. Induction.
3. The contact of dissimilar metals.
4. The contact of metals with liquids.
5. "A mere variation of the character of the contact of two bodies."
6. "Chemical action produces a continuous flow of electricity (voltaic electricity)."
7. "Heat, suitably applied to dissimilar metals, produces a continuous flow of electricity (thermo-electricity)."
8. The heating and cooling of certain crystals (pyro-electricity)."
9. "The motion of magnets and of bodies carrying electric currents (magneto-electricity)."
10. "The friction of sand against a metal plate."
11. "The friction of condensed water particles against a safety valve, or, better still, against a boxwood muzzle, through which steam is driven (Armstrong's hydro-electric machine)." (*Tynd*. 110.)

(*b*) *Electrical Measurement.*

(6.) In electro-static measurement, it is found by experiment that when F is the force of repulsion or attraction between two small electrified bodies, Q and Q_1 the charges or quantities in those bodies, and D their distance apart, then

$$F = \frac{Q\,Q_1}{D^2}, \text{ and, where } Q_1 = Q,\ F = \frac{Q^2}{D^2},$$

whence the value of Q can be determined. (*Jen*. 95.)

(7.) If Q is the quantity of electricity on a body,
    S ,, ,, capacity of such body, and
    P ,, ,, potential ,, ,,
then Q = PS. (*Jen*. 96 & 97.)

(8.) "The following table of dimensions and constants is taken from the British Association Report on Electrical Standards, 1863."

"Fundamental units—
    Length = L.
    Time = T.
    Mass = M."

NOTES ON LIGHTNING.   3

I. A c 9—12.

"Derived mechanical units—

$$\text{Work} = W = \frac{L^2 M}{T^2}.$$

$$\text{Force} = F = \frac{L M}{T^2}.$$

$$\text{Velocity} = V = \frac{L}{T}."$$

"Electrostatic system of units—

$$\text{Quantity of electricity} = Q = \frac{L^{3/2} M^{1/2}}{T}.$$

$$\text{Strength of electric current} = C = \frac{L^{3/2} M^{1/2}}{T^2}.$$

$$\text{Electro-motive force} = E = \frac{L^{1/2} M^{1/2}}{T}.$$

$$\text{Resistance of conductor} = R = \frac{T}{L}."$$

(*Jen.* 163 & 164.)

(9.) Distance $= l$. Mass $= m$. Time $= t$. Velocity $= \frac{l}{t}$. Acceleration $= \frac{v}{t} = \frac{l}{t^2}$. Force = acceleration × mass $= \frac{m l}{t^2}$. Work = force × distance $= \frac{m l^2}{t^2}$. (*Desch.*)

(10.) Quantity $= f^{\frac{1}{2}} l = \frac{m^{\frac{1}{2}} l^{\frac{3}{2}}}{t}$. Potential $= \frac{\text{work}}{\text{quantity}}$ $= \frac{m^{\frac{1}{2}} l^{\frac{1}{2}}}{t}$. Capacity $= \frac{\text{quantity}}{\text{potential}} = l$. (*Desch.*)

(11.) Ohm's law.

$$\text{Current} = \frac{\text{Electro-motive force}}{\text{Resistance}}. \quad (\textit{Jen. } 82.)$$

(*c*) *Potential.*

(12.) Potential may be compared with a head of water. (*Jen.* 10 & 40.)

I. A *d e* 13—20.

(13.) "Difference of potential is a difference of electrical condition in virtue of which work is done by positive electricity in moving from the point at a higher potential to that at a lower." (*Jen.* 26.)

(*d*) *Insulation.*

(14.) An insulated surface can only lose its electricity gradually. (*Jen.* 3.)

(15.) "We may expect that if from any cause the distribution of electricity in a body can be varied, even without its total amount being changed, this redistribution will take place almost instantaneously in the electrified conductor, and much more slowly in the electrified insulator." (*Jen.* 3.)

(16.) "No substance is found to insulate so perfectly as to possess the power of keeping the two electricities asunder for more than a limited time. A perpetual leakage is always occurring from the one to the other through the mass of the insulator, until the combination or neutralisation is complete, and all signs of electricity disappear." (*Jen.* 8.)

(*e*) *Induction.*

(17.) If a positively charged body A be brought near an uncharged body B, "it attracts negative electricity to that end of the body B which is near it, and repels positive electricity to the remoter portions of B." (*Jen.* 11.)

(18.) "Induction of electricity must take place in the space surrounding every electrified body." (*Jen.* 12.)

(19.) "Induction always takes place between two conductors at different potentials separated by an insulator." (*Jen.* 13.)

(20.) "The very existence of the original charge implies the induced charge." (*Jen.* 13.)

I. A ƒ 20a—26.

(20a.) "The only manner in which we can in any way account for the observed facts of attraction, repulsion, and induction is by assuming that the forces are transmitted by a strain or distortion of the medium which fills the space between the electrified bodies." (*Gord.* i. 20.)

(ƒ) *Condensation.*

(21.) "Whenever a conductor is charged, a kind of Leyden jar is necessarily formed. The conductor is the inner coating, the air the dielectric, and the nearest surrounding conductors . . . form the outer coating." (*Jen.* 19.)

(22.) "A condenser is an apparatus for condensing a large quantity of electricity on a comparatively small surface. The form may vary considerably, but in all cases consists essentially of two insulated conductors, separated by a non-conductor, and depends on the action of induction." (*Gan.* 622.)

(23.) Condensation is the "operation of obtaining electricity of high potential from a source of comparatively low potential." (*Chath.* 134.)

(24.) Franklin found that the electricity in a condenser resides in the dielectric. This can be proved by making the coatings of a Leyden jar movable. (*Tynd.* 78.)

(25.) "The coefficient by which the capacity of an air condenser must be multiplied in order to give the capacity of the same condenser when another dielectric is substituted for air is constant for each substance, and is called the 'specific inductive capacity' of the dielectric." (*Jen.* 97.)

(25a.) Specific inductive capacity is "the specific power of the substance of which the insulator is composed of receiving and transmitting that electric strain which we call induction." (*Gord.* i. 69.)

(26.) Approximate specific inductive capacities:—

I. A *g* 27—31; B 1, 2.

Air	1	India-rubber	2·8
Pitch	1·8	Gutta-percha	4·2
Glass	1·9	Mica	5

(*Jen.* 96.)

(*g*) *Electric Sparks.*

(27.) Electric sparks are said to overcome the resistance of the air, but this resistance has nothing in common with the resistance which is the subject of Ohm's law." (*Jen.* 92.)

(28.) Sir Charles Wheatstone found in a special case the duration of an electric spark to be $\frac{1}{24000}$ of a second, but this was the maximum. In other cases it was less than $\frac{1}{1000000}$ of a second. (*Tynd.* 85.)

(29.) "The recombination of the two electricities which constitute the electrical discharge may be either continuous or sudden; . . . . sudden, as when the opposite electricities accumulate on the surface of two adjacent conductors, till their mutual attraction is strong enough to overcome the intervening resistances, whatever they may be." (*Gan.* 638.)

(30.) "The brush forms when the electricity leaves the conductor in a continuous flow, the spark when the discharge is discontinuous." (*Gan.* 639.)

(31.) "When an electric discharge is sent through gunpowder placed on the table of a Henley's discharger, it is not ignited, but is projected in all directions." (*Gan.* 643.)

(B.) THE CONDUCTIVITY OF MATERIALS.

(1.) Cavendish estimated iron to be 400,000,000 times more conductive than water. (*Harr.* 7.)

(2.) Relative conductivity of building metals (according to Sir Wm. Snow Harris):—

Lead	1	Zinc	4
Tin	2	Copper	12
Iron	2½		

(*W. O.* 1875, 36.)

(3.) Extracts from table of conductivity given by Sir W. S. Harris:—

	"Conductors."		"Insulators."
"Most Perfect."	Metals. Charcoal. Plumbago. Flame. Smoke.	"Less Perfect."	Ice at 0°. Dried vegetable substances. Dried animal substances. Parchment, leather, feathers. Bituminous matter. Silk.
"Less Perfect."	Living animals. Living vegetables.		
"Imperfect."	Wood. Snow and Ice. Aqueous vapour. Common earth and stone. Dry chalk and lime. Marble and porcelain. Paper.	"Most Perfect."	Animal fur and hair. Dry gases, including atmosphere. Pure steam of high elasticity. Glass and all vitrefactions. All resins and resinous bodies.

(*Harr.* 7.)

(3*a*.) Matthiesen gives graphite and gas-coke as 1,450 to 40,000 times less conductive than pure copper. (*Jen.* 257.)

(4.) Rarefied gases are found to be tolerably good conductors. (*Jen.* 93.)

(5.) The conductivity of silver to that of gutta-percha is 85×10^{20} to 1. (*Jen.* 85.)

(6.) Extracts from Roget's "Catalogue of Bodies in the order of their Conducting Power:"—

The least perfect or least oxidable metals.	Living animals.
The more oxidable metals.	Flame.
Charcoal prepared from the harder woods and well burned.	Smoke.
Plumbago.	Steam.
Metallic ores.	Rarefied air.
Animal fluids.	Earths and stones in their ordinary state.
Pure water.	Vegetable ashes.
Snow.	Animal ashes.
Living vegetables.	Ice below 13° Fahrenheit.
	Lime.

I. B 7—10.

Chalk.	Cotton.
India-rubber.	Feathers.
Siliceous and argillaceous stones in proportion to their hardness.	Hair, especially that of a living cat.
	Wool.
	Dyed silk.
Dry marble.	Bleached silk.
Porcelain.	Raw silk.
Dry atmospheric air and other gases.	Glass and other vitrefactions.
	Fat.
Leather.	Resins and bituminous substances.
Dry paper.	(*Rog.* ii. 6.)

(7.) The deposition of moisture on an insulator increases the conducting power. (*Rog.* ii. 7.)

(8.) Lunn gives practically the same order of conductivity as Roget. The following are the principal differences, viz. :—

(a) In lieu of "pure water."
 1st. "Sea water."
 2nd. "Spring water."
 3rd. "Rain water."

(β) Between "rain water" and "snow," "ice above 13° Fahr." is inserted.

(γ) For "earths and stones in their ordinary state"—"moist earths and stones."

(δ) Between "porcelain" and "air," "dry vegetable bodies" are inserted. (*Lunn*, 72.)

(9.) Extracts from Tyndall's list of conductivity :—

"*Conductors.*"	"*Insulators.*"
The common metals.	Chalk.
Well-burned charcoal.	India-rubber.
Rain water.	Paper.
Linen.	Hair.
Vegetables and animals.	Silk.
"*Semi-Conductors.*"	Glass.
Wood.	
Marble.	
Straw.	(*Tynd.* 18.)

(10.) Conductivity of pure copper = 100,000,000. Ditto of solution of concentrated common salt = 31·52. (*Clark.*)

I. B 11—14.

(11.) Extracts from list in R. E. Aide-Mémoire of "Bodies arranged in the order of their relative conducting power:"—

"Conductors."	*"Non-conductors or Insulators."*
"Most Perfect."	*"Less Perfect."*
All known metals.	Ice below 0° of Fahrenheit.
Well-burned charcoal.	Dried vegetable substances.
Plumbago.	Dried animal substances.
Burning gaseous matter, as flame.	Parchment, leather, feathers.
Smoke.	Fur and hair.
"Less Perfect."	Silk.
Dry chalk and lime.	*"Most Perfect."*
Marble and porcelain.	Dry air and other gases.
Paper.	Pure steam of high elasticity.
Wood in its ordinary state.	Glass and all vitrefactions.
Water.	All resins and resinous bodies.
Snow and ice from 32° to 0°.	(*R. E. A.* 56.)
Living animals.	
Living vegetables.	
Aqueous vapour.	
Common earth and stone.	

(12.) *Order of conductivity of metals, &c.:—*

Silver	100	Tin	13·1
Copper	99·9	Lead	8·3
Gold	80	German silver	7·7
Zinc	29	Mercury	1·6
Platina	18	Bismuth	1·2
Iron	16·8	Graphite	0·07

(*R. E. A.* 56.)

(13.) Those bodies are "conveniently designated as conductors which when applied to a charged electroscope discharge it almost instantaneously, semi-conductors being those which discharge it in a short but measurable time—a few seconds for instance; while non-conductors effect no discharge in the course of a minute." (*Gan.* 587.)

(14.) The following list "is arranged in order of decreasing conductivity, or, what is the same thing, of increased resistance. The arrangement is not invariable, however. Conductivity depends on many physical conditions." The following are extracts from the list:—

I. B 15, 16. C a 1—3.

"Conductors."	"Semi-conductors."	"Non-conductors."
Metals.	Wood.	Ice at 25° C.
Well-burnt charcoal.	Ice at 0°.	Lime.
Graphite.		Caoutchouc.
Aqueous solutions.		Air and dry gases.
Water.		Dry paper.
Snow.		Silk.
Vegetables.		Glass.
Animals.		Resins.
Linen.		
Cotton.		(*Gan.* 586.)

(15.) The average result of the researches of Sir Humphrey Davy, Becquerel, Lenz, Ohm, and Pouillet as regards the relative conductivity of copper and iron fixes the proportion as that of 100 to $16\frac{1}{2}$. (*And.* 55.)

(16.) Specific resistances of substances in absolute units, as given by Professor Everett, F.R.S., in his "Units and Physical Constants:"—

Silver, hard drawn 1609	Lead, pressed 19,847
Copper „ 1642	Water at 22° C. . . $7 \cdot 18 \times 10^{10}$
Gold „ 2154	Gutta-percha at 24° C. $3 \cdot 53 \times 10^{23}$
Iron, annealed 9827	(*Gord.* i. 259.)

(C.) THE ELECTRICITY AND MAGNETISM OF THE EARTH.

(*a*) *Atmospheric Electricity.*

(1.) "The electric telegraph forces us to combine our ideas with reference to terrestrial magnetism and atmospheric electricity. We must look on the earth and air as a whole—a globe of earth and air—and consider its electricity, whether at rest or in motion." This science is that of "terrestrial electricity." (*Thoms. Tel.* 15/1/74.)

(2.) "The subject of atmospheric electricity is yet in its infancy, and is one of extreme difficulty." (*Do. S. T. E.* 369.)

(3.) "No connection between atmospheric electricity, thunderstorms, or generally the state of the weather, has yet been discovered." (*Do.*)

I. C 4—14.

(4.) "There is no reason to suppose that clouds are essential to electrical discharge in the atmosphere. On the contrary, instances are recorded, both in ancient and modern times, of lightning flashes occurring in a perfectly clear sky." (*Do.*)

(5.) "In fair weather, the surface of the earth is always, in these countries at all events, found negatively electrified." (*Do. Tel.* 15/1/74.)

(6.) "Positive electricity of the air is merely inferential. The result obtained in daily observations is precisely the same as if the earth were electrified negatively, and the air had no electricity in it whatever." (*Do.*)

(7.) "Probably all space is non-conductive, and the upper regions of the air have no electricity." (*Do.*)

(8.) The atmosphere always contains free electricity, sometimes positive, and sometimes negative. (*Gan.* 827.)

(9.) The electricity of the ground is always negative; but it varies according to the hygrometric and thermometric states of the air. (*Do.* 828.)

(10.) "Many hypotheses have been propounded to explain the origin of atmospheric electricity. Some have ascribed it to the friction of the air against the ground, some to the vegetation of plants, or to the evaporation of water. Some again have compared the earth to a vast voltaic pile, and others to a thermo-electrical apparatus. Many of these causes may in fact concur in producing the phenomena." (*Do.*)

(11.) Volta showed that evaporation produced electricity, and Pouillet and others have proved that the evaporation must be that of undistilled water. (*Do.*)

(12.) Evaporation and vegetation are great sources of the electricity of the atmosphere. (*Hersch.* 127.)

(13.) Friction is probably one of the causes of atmospheric electricity. Evaporation is a more powerful source, but it must be accompanied by chemical decomposition. Combustion and vegetation are also sources. (*Kaem.* 336.)

(14.) "It is considered that electricity is being per-

I. C *b* 15—22.

petually evolved from the earth (as from a huge electric machine) by the incessant changes in the mechanical as well as chemical condition of its constituents; such changes for instance as those accompanying variations of temperature produced by the enormous extent of evaporation from the land and fresh water as well as from the ocean—by the absorption and re-irradiation of solar heat; by the escape of central heat; or by the decomposition and recomposition perpetually in progress over the face of the earth, of all descriptions, from slow putrefaction to rapid combustion, &c., &c., all of which are more or less associated with changes in electric condition." (*Nels. A. M.*)

(*b*) *Rain and Hail.*

(15.) D'Alibard and Franklin found that every shower of rain is accompanied by electricity. (*Kaem.* 328.)

(16.) Rain and (especially) hail are probably causes of lightning. The electriferous globules of the clouds coalesce into rain, and a sudden increase of electric tension results. Each great flash of lightning is generally succeeded by a sudden rush of rain. (*Hersch.* 131.)

(17.) The condensed vapours liberate electricity. Rain or hail follows flashes of lightning. These are the effect of rain rather than the cause of it. (*Kaem.* 368.)

(18.) Rain falling during thunderstorms contains nitric acid. (*Ar.* 64.)

(19.) Thundery weather is known to turn milk sour, to spoil beer, and to hasten the corruption of meat. (*Do.* 98.)

(20.) "The formation of hail appears to be indisputably connected with the presence of an abundant quantity of fulminating matter in the clouds." (*Do.* 235.)

(21.) Hail commits great ravages in France on agriculture, and especially in the vine districts. (*Do.* 233.)

(22.) The use of pointed captive balloons is suggested by Arago in order to dissipate hailstorms and thunderstorms. (*Do.*)

I. C *c* 23—32.

(23.) "Hailstorms are those in which the development of electricity attains the largest proportions. The thick clouds in which the meteor becomes elaborated are laden with a large quantity of electrical fluid." (*Flam.* 442.)

(24.) Hail is formed of globules of ice, and generally precedes thunderstorms. It is not well accounted for. (*Gan.* 816—824.)

(25.) It generally falls in the hottest time of day, and in spring or summer. (*Do.*)

(26.) Electricity is present in dew, fog, and snow. (*Kaem.* 342.)

(*c*) *Clouds.*

(27.) Cloud is only fog. Fog is always in a comparatively high electric state. (*Hersch.* 129.)

(28.) The following heights of thunderclouds have been observed :—

By De L'Isle, at Paris, in 1712	26,510 feet.
„ Abbé Chappe, at Tobolsk, in 1761	10,960 „
„ Lambert, at Berlin, 1773	{ 6,234 / 5,230 } „
„ Le Gentil, at Mauritius	2,953 „
„ „ Pondicherry, in 1769	10,827 „
„ D'Abbadie, at Abyssinia, in 1843—5	{ 696 / 6,680 } „
Once in Austria (See incident No. 3, Chap. III.)	92 „

(*Ar.* 17.)

(29.) Thunderclouds are about 1,300 to 1,400 yards distant from the earth in winter, and 3,300 to 4,400 in summer. (*Gan.* 816—824.)

(30.) The thickness of a low-lying stormcloud at Gratz, in Austria, on 15th June, 1826, was found to be 120 feet. (*And.* 69.)

(31.) Clouds approach each other either by electric action, or when driven by winds. (*Harr.* 59.)

(32.) Wind and electric attraction tend to make the clouds approach the earth. (*Mann.* 1875, 531.)

I. C *d* 33—41.

(33.) When a charged thundercloud " is generated in the upper regions of the air, and hemmed round by its incumscribing insulation, all the complicated phenomena of induction immediately appear." (*Do.* 530.)

(34.) "The broad masses of insulated clouds are conductors ready to receive large charges . . . . the surrounding spaces of clearer and drier air are the insulators that imprison the accumulating charge, and the vapours that ascend from the earth and drift in with the winds from side regions are the carriers and feeders of the charge." (*Do.*)

(35.) "The real function of the cloud is simply the bringing into continuous electrical communication a wide stretch of electrically charged air." (*Do.*)

(36.) "When a stormcloud hangs low over the earth, the negative reaction of that part of the ground is very largely intensified by induction." (*Do.*)

(*d*) *Terrestrial Electricity.*

(37.) "When the atmosphere is tempestuous, there are simultaneously great perturbations in the interior of the earth, and at the surface, or below the surface of waters." (*Ar.* 93.)

(38.) Cases have happened during thunderous weather of surfaces of springs and wells being troubled, of inundations being caused, and of subterraneous noises. (*Do.* 94.)

(39.) By the influence of a thunderstorm, flames may be developed under water, and shoot upwards from its surface. (*Do.* 100.)

(40.) It cannot be doubtful that local circumstances influence the frequency of thunderstorms. (*Do.* 115.)

(41.) Mr. L. W. Dillwyn, writing in 1803, considers that limestone regions are more subject to thunderstorms than others. (*Do.* 116.)

I. C 42—51.

(42.) The course of lightning is influenced by the terrestrial bodies near which it explodes. (*Do.* 92.)

(43.) Thunderclouds often follow the course of rivers. (*Do.* 7.)

(44.) "There can be no doubt that thunderstorms will visit some districts in preference to others, and that lightning will descend constantly on some selected spots, and will entirely keep away from others." (*And.* 63.)

(45.) "There can be no doubt, from thousands of observations made, that it is one of the characteristics of the electric force to seek its way towards waters." (*Do.* 69.)

(46.) The surface of the earth is one of the terminating planes of the electrical action. (*W. O.* 1858. *App. B.* 4.)

(47.) The terminating electric plane of a lightning discharge is sometimes beneath the surface of the ground. (*Do.* 1875, 2.)

(48.) Light dry soils, such as shingle and sand, are to be regarded as non-conducting matter resting on the electric surface. (*Do.* 3.)

(49.) The electricity on the earth's surface is induced by the clouds. (*Harr.* 11 *et passim.*)

(50.) "The potential of the earth's surface is assumed as the zero or datum level from which all other potentials are measured; nevertheless we know that the potentials of different places on and in the earth differ considerably, sometimes to the extent of several hundred volts, though this is rare. We obtain this information from the currents observed to flow through wires joining parts of the earth widely separated." (*Jen.* 365.)

(51.) "Not much is known of the distribution of electricity on the surface of the earth. According to Sir Wm. Thomson the most probable distribution is analogous to that which would be produced if the earth's surface generally were charged with negative electricity held as a charge on the inner armature of a condenser, the outer armature of which was in the upper regions of the atmosphere, the lower part of which acts as a dielectric." (*Do.*)

I. C *e* 52—60.

(52.) The earth has been named the common reservoir of electricity. (*Gan.* 58.)

(53.) "Peltier drew the conclusion from his experiments 'that the earth itself, and more particularly the fiery liquid mass forming the inner bulk of it, over which the solid crust and the ocean lie, both thinner in comparison than the skin of an apple, forms one immense reservoir of electricity.' As light comes from the sun, generated as we believe by heat, so the electric force, he held, comes from the interior of the globe, likewise generated by heat." (*And.* 71.)

(54.) "On the whole, Peltier's explanation, such as it is, may fairly be accepted, in the present state of scientific investigation, as one of the best that can be given." (*Do.* 72.)

(*e*) *Earth Currents.*

(55.) Earth currents are irregular currents of electricity which manifest themselves by abnormal disturbances of telegraphs. (*Gan.* 837.)

(56.) Sabine considers earth currents are magnetic disturbances "due to a peculiar action of the sun, and probably independent of its radiant heat and light." (*Do.*)

(57.) Balfour Stewart considers earth currents as secondary currents due to small but rapid changes in the earth's magnetism. (*Do.*)

(58.) Earth currents are constantly passing through the telegraph wires, "due to the varying potential of different parts of the earth's surface." (*Cull.* 28.)

(59.) Earth currents depend "either on difference of potential between the earth at the two stations or on induction from passing clouds." (*Jen.* 310.)

(60.) Powerful earth currents were experienced on telegraph lines in Ireland and Scotland on August 12th, 1880. On the same day there was a thunderstorm in the S.W. of Ireland, and on the night of that day a brilliant aurora was seen in Aberdeen and other parts of Scotland, and a violent thunderstorm occurred at Vienna. (*Stand.* 14/8/80. *Tel.* 1/9/80.)

(60*a*.) "Magnetic observations are complicated by the existence of certain currents of electricity which move in the earth." (*Gord.* i. 199.)

(60*b*.) Mr. C. V. Walker, F.R.S., published in 1861—2 a series of investigations on earth currents, and determined therefrom :—

1st. "That currents of electricity are at all times moving in definite directions in the earth."

2nd. "That their direction is not determined by local causes."

3rd. "That there is no apparent difference, except in degree, between the currents collected in times of great magnetic disturbance, and those collected during the ordinary calm periods."

4th. "That the prevailing directions of earth currents or the currents of most frequent occurrence are approximately N.E. and S.W. respectively." (*Do.*)

(*f*) *The Earth's Magnetism.*

(61.) "Magnets are substances which have the property of attracting iron, and the term magnetism is applied to the cause of this attraction, and to the resulting phenomena." (*Gan.* 557.)

(62.) "Magnetic substances are substances which, like iron, steel, and nickel, are attracted by the magnet." (*Do.* 562.)

(63.) "The earth acts as a great magnet. Dr. Gilbert, of Colchester, made that clear nearly three hundred years ago." (*Thoms. Tel.* 15/1/74.)

(64.) "The declination is accidentally disturbed in its daily variations by many causes, such as earthquakes, the aurora borealis, and volcanic eruptions. The effect of the

I. C 65—69.

aurora is felt at great distances. Auroras which are only visible in the north of Europe act on the needle even in these (Paris) latitudes, where accidental variations of 20′ have been observed. In polar regions the needle frequently oscillates several degrees; its irregularity on the day before the aurora borealis is a presage of the occurrence of this phenomenon." (*Gan.* 567.)

(65.) "Another remarkable phenomenon is the miscellaneous occurrence of magnetic perturbations in very distant countries. Thus Sabine mentions a magnetic disturbance which was felt simultaneously at Toronto, the Cape, Prague, and Van Diemen's Land. Such simultaneous perturbations have received the name of magnetic storms." (*Do.*)

(66.) "The magnetic intensity increases with the latitude. Humboldt found a point of minimum intensity on the magnetic equator in Northern Peru." (*Do.* 573.)

(67.) "Observations have led to the discovery that the magnetism of the earth is in a state of constant fluctuation like the waves of the sea." (*Do.* 574.)

(68.) "Ampère assumes that each individual molecule of a magnetic substance is traversed by a closed electric current." .... "The resultant of the actions of all the molecular currents is equivalent to that of a single current which traverses the outside of a magnet." (*Do.* 719.)

(69.) "In order to explain on this supposition terrestrial magnetic effects, the existence of electrical currents is assumed which continually circulate round our globe from east to west perpendicular to the magnetic meridian. The resultant of their action is a single current traversing the magnetic equator from east to west. These currents are supposed to be thermo-electric currents due to the variations of temperature caused by the successive influence of the sun on the different parts of the globe from east to west. These currents direct magnetic needles and impart a natural magnetisation to iron minerals." (*Do.*)

(70.) As regards the distribution of the magnetic force on the surface of the earth, "it is conceivable that this force may be wholly due to currents flowing round the earth, and maintained by the thermo-electric action of the sun, or to some other cause connected with the rotation of the earth." (*Jen.* 367.)

(71.) "It is one of the greatest mysteries of science. .... What can be the cause of this magnetism in the interior of the earth? .... Electric currents afford the more favourable hypothesis. ... But what sustains the electric currents? .... We have none of the elements of the problem of thermo-electricity in the underground temperature which could possibly explain, in accordance with any knowledge of thermo-electricity, how there could be sustained currents round the earth." (*Thoms. Tel.* 15/1/74.)

(72.) "It was suggested by the great astronomer (Halley) that there is a nucleus in the interior of the earth, and that the mystery is explained by a magnet not rigidly connected with the upper crust of the earth, but revolving round an axis of figure of the outer crust, and exhibiting a gradual precessional motion independent of the precessional motion of the outer rigid crust." (*Do.*)

(73.) "If we could have simultaneous observations of the underground currents, of the three magnetic elements, and of the aurora, we should have a mass of evidence from which, I believe, without fail, we ought to be able to conclude an answer more or less definite to the question I have put." (*Do.*)

(73*a.*) "There is no doubt that the earth is affected by electrical phenomena occurring in the sun." (*Gord.* i. 21.)

(73*b.*) Professor Balfour Stewart has pointed out that the magnetic storm of unprecedented magnitude that occurred in August and September, 1859, was synchronous with the period of maximum activity of one of the largest sun-spots ever observed. (*Do.* 197.)

I. C*g* 74—82.

(*g*) *Auroræ.*

(74.) Aurora is "a remarkable luminous phenomenon, which is frequently seen in the atmosphere at the two terrestrial poles." (*Gan.* 837.)

(75.) "The constant direction of their arc as regards the magnetic meridian, and their action on the magnetic needle show that they (auroræ) ought to be attributed to electrical currents in the higher regions of the atmosphere." (*Do.*)

(76.) "This hypothesis is confirmed by the circumstance observed in France and other countries on August 29th and September 1st, 1859, that two brilliant auroræ boreales acted powerfully on the wire of the electric telegraph." (*Do.*)

(77.) "According to M. De la Rive, the auroræ boreales are due to electric discharges which take place in Polar regions between the positive electricity of the atmosphere and the negative electricity of the terrestrial globe; electricities which themselves are separated by the action of the sun on the equatorial regions." (*Do.*)

(78.) "The aurora borealis is one of the grand results of atmospheric electricity. Instead of a furious and violent storm limited to a few leagues, it is a gentle and gradual recomposition of the negative fluid of the earth with the positive fluid of the atmosphere taking place in the aërial heights, in the upper hydrogenous atmosphere. This disengagement of electricity in a vast sheet is only visible at night." (*Flam.* 497.)

(79.) "We have the strongest possible reason for believing that aurora consists of electric currents." (*Thoms. Tel.* 15/1/74.)

(80.) "Aurora borealis is properly comparable with the phenomenon presented by vacuum tubes." (*Do.*)

(81.) Magnetic storms are always associated with auroræ. (*Do.*)

(82.) "According to Balfour Stewart, auroræ and earth currents are to be regarded as secondary currents due to small but rapid changes of the earth's magnetism; he likens the body of the earth to the magnetic core of a Ruhmkorff's machine, the lower strata of the atmosphere forming the insulator, while the upper and rarer, and therefore electrically conducting, strata may be considered as the secondary coil. On this analogy the sun may perhaps be likened to the primary current which performs the part of producing changes in the magnetic state of the core." (*Gan.* 837.)

(83.) "It is principally in the neighbourhood of the Polar circle, where thunderstorms are rare, that these manifestations of terrestrial electricity are seen to the fullest advantage." (*Flam.* 497.)

(84.) "A French scientific commission to the North observed one hundred and fifty auroræ in two hundred days." (*Gan.* 836.)

(85.) "They are visible at a considerable distance from the Poles and over an immense area. Sometimes the same aurora borealis has been seen at the same time at Moscow, Warsaw, Rome, and Cadiz." (*Do.* 837.)

(*h*) *St. Elmo's Fires.*

(86.) In thunderstorms the projecting parts of bodies, and especially the metallic parts, sometimes shine with rather a vivid light. This light is called St. Elmo's Fire. (*Ar.* 102.)

(87.) "The St. Elmo's Fires are a slow manifestation of electricity, a quiet and steady outflow which radiates gently over the topmost parts of lightning conductors of buildings and vessels during thunder weather, when the terrestrial electric tension is strongly attracted by that of the clouds." (*Flam.* 493.)

(88.) "The 'electric glow' is from a negatively electrified pointed conductor." (*Tynd.* 92.)

(89.) The glow is sometimes seen on the masts of ships, and it is mentioned by the ancients as appearing on the points of lances. It is called St. Ermo's, or St. Elmo's Fire, after the sailor's saint, Erasmus, who suffered martyrdom at Gaeta at the beginning of the fourth century." (*Do.*)

(90.) Nitrogen gas has the greatest power of originating this luminous appearance when it is artificially formed. (*Harr.* 45.)

(91.) St. Elmo's Fires are several times mentioned by Pliny, Seneca, Cæsar, and Livy. English sailors call them "comazants." (*Do.* 19.)

(92.) In January, 1748, several comazants settled on the spindles of the masts of the merchant ship *Dover*, and burnt like large torches. (*Do.*)

(93.) In May, 1821, in the North Atlantic Ocean, between Bermuda and Halifax, St. Elmo's Fires were observed on the masts of H.M.S. *Newcastle*. (*Do.*)

(94.) Priestley's "History of Electricity" states that the cross of the church steeple of Plauzet, in France, always during thunderstorms had its three pointed extremities surrounded with a body of flame. (*Do.*)

(95.) St. Elmo's Fires are often seen over the spires of Notre Dame, Paris, during violent thunderstorms. (*Flam.* 494.)

(96.) On the 2nd March, 1869, the cross on the steeple (130 feet high) of the Church of St. Catherine de Fierbois, Chinon, France, appeared, towards the end of a storm, with "a crown of fire" around it. No thunder had been audible. (*Do.*)

(*i*) *Waterspouts.*

(97.) Waterspouts "are masses of vapour suspended in the lower layers of the atmosphere, which they traverse, and endowed with a gyratory motion." (*Gan.* 819.)

(98.) Waterspouts are generally accompanied by hail and rain, and often emit thunder and lightning. (*Do.*)

(99.) "When they take place on the sea .... the water is disturbed and rises in the form of a cone, whilst the clouds are depressed in the form of an inverted cone; the two cones then unite to form a continuous column from the sea to the cloud." (*Do.*)

(100.) The origin of waterspouts is not known. "Peltier and many others ascribe to them an electrical origin." (*Do.*)

(101.) At Illinois, U.S., on the 4th of June, 1814, Mr. Griswold observed a waterspout with incessant lightning between the clouds and the earth near the surface of the spout; but no thunder was heard. (*Ar.* 155.)

(102.) At Banbury, in England, on the 3rd of December, 1872, what was at first supposed to be a "fireball" was seen by four persons from different points. It appears to have been a whirlwind of vapour or waterspout, accompanied by a whizzing sound, sulphurous smell, and sparks or flashes, rising high into the air, and passing about six to ten feet over the ground for about two miles, uprooting trees on its way, till, at a place where the ground had the appearance of being cut up by a cannon ball, it vanished. (*S. T. E.* 371.)

(103.) A waterspout was seen on the rifle ranges at Norwich in July, 1880. It lasted half an hour. (*Graph.* 24/7/80.)

(104.) A waterspout passed over Gower, in Glamorganshire, on the 22nd July, 1880. Thunderclouds had previously gathered. It consisted of a large slender column of dense vapour or rain descending from the clouds to the earth, and lasted about five minutes, when it was absorbed in heavy black clouds, and these then discharged for about an hour a perfect deluge of rain. (*S. M. M. August*, 1880.)

(*k*) *Earthquakes.*

(105.) The proximate cause of earthquakes "seems referable to the action of internal heat or fire." (*Enc. Br. Art. Earthquake.*)

I. C 106—112.

(106.) "There is good reason for holding that earthquakes are closely connected with volcanic agency. Both probably spring from the same cause." (*Do.*)

(107.) "Whilst the action of water has been generally to wear down, transport, and redeposit in nearly regular and horizontal order, or in other words, to restore the level of the earth's surface, the action of heat has, in conjunction with electricity, &c., tended to disturb that level, and to raise some portions of the surface above others." (*Portl. A.M.*)

(108.) "The facts of disturbance are palpable, and the nature of the forces producing them can be inferred by reasoning, though not demonstrated by observation. Electricity may be fairly classed with such forces, and yet it may only be a secondary cause; but assuredly ample reason has been adduced for the assumption that heat must at least be a primary one." (*Do.*)

(109.) "Mr. Howorth quotes from Dr. Zollner's paper in the Philosophical Magazine on the subject of the correlation of earthquakes with magnetic disturbance, wherein it is stated that Kriel has given many cases where magnetic disturbances coincide with earthquakes; hence he thinks connection between the two phenomena probable." (*Tel.* 15/9/74.)

(110.) In 1822, there occurred in France, and over a great part of the Continent, "an extraordinary number of violent thunderstorms, accompanied by earthquakes and simultaneous eruptions of Mount Vesuvius, the latter on a scale not witnessed for centuries." (*And.* 76.)

(110*a*.) A great earthquake occurred in Chili in 1822. (*Portl. A.M.*)

(111.) At the earthquake or upheaval of Sabrina, a small island among the Azores, in 1811, Captain Tillard observed columns of dust and ashes and much lightning. (*Ar.* 12.)

(112.) Cuvier asserts that "both men and animals suffer from a certain *malaise* during the period of these convul-

I. C *l* 113—117.

sions, and he attributes the reason to electrical disturbance." (*Stand.* 12/11/80.)

(113.) Earthquakes occurred in 1880 as follows: At Manilla in July, at Smyrna in July, at Valparaiso in September, at Lisbon, Madrid, and other parts of the peninsula in October, at Agram and over Southern Austria in November and December, and at Odessa, Bucharest, and over South Eastern Europe in December. (*Stand. July—Dec.* '80.)

(113*a*.) The shocks of earthquake at Agram in January, 1881, were accompanied by "loud subterranean thunder." (*Stand.* 27/1/81.)

(*l*) *Volcanic Eruptions.*

(114.) Pliny wrote to Tacitus, A.D. 79, of lightning attending an eruption of Mount Vesuvius. Padre della Torre, in 1182, mentions lightnings as often seen there in the midst of smoke. Bracini relates of the eruption of Vesuvius in 1631 that frequent lightnings issued from a cloud of smoke from the crater and killed several persons. Valetta mentions lightnings happening at the eruption of 1707. The same circumstances are recorded of the eruption of 1767. (*Ar.* 11.)

(115.) Sir William Hamilton mentions lightning without thunder at the eruption of Vesuvius of 1779, and with thunder at that of 1794. A house struck at San Jorio by the lightning during the latter eruption showed the same effects as if occasioned by atmospheric lightning. (*Do.* 12.)

(116.) Earthquakes are a portion of the phenomena of volcanic eruptions, "the earthquake often preceding the volcanic eruption, being the result of unusual movements in the interior of the earth, which is connected with its surface by the volcano." (*Portl. A. M.*)

(117.) "The phenomena more nearly preceding an eruption (of Mount Vesuvius) are the occurrence of earthquakes, increasing in intensity and frequency for some

I. C 118—122; D a 1—6.

days beforehand, also the irregularity of the diurnal variations of the magnetic needle." (*Fras. R. E. P.*)

(118.) "One of the remarkable attendants of an eruption is the frequency of lightning flashes, considered by Dr. Palmieri to be due to the condensation of the vapour of water from the crater; just as in an ordinary thunderstorm lightning occurs at the time the vapour is condensing, as is proved by the rain that follows." (*Do.*)

(119.) Seneca mentions lightning appearing at an eruption of Mount Etna. In 1755 lightning attended an eruption of Mount Etna. (*Ar.* 12.)

(120.) At a little volcano which appeared in 1831 between Sicily and Pantellaria, columns of black dust arose with lightning and thunder. (*Do.* 13.)

(121.) During the earthquakes at Manilla in July, 1880, all the volcanoes of the island were in eruption. (*Stand. July*, 1880.)

(122.) Mount Vesuvius was in eruption in September and November, 1880. (*Do.* 9/80 & 11/80.)

(D.) LIGHTNING DISCHARGES.

(a) *The Nature of Lightning.*

(1.) A thunderbolt is defined as "a shaft of lightning, particularly if passing in a direction towards the earth." (*Chamb.*)

(2.) Seneca defined thunderbolts as lightnings that reach the ground. (*Ar.* 24.)

(3.) *Foudre* (*anglice*, thunderbolt), is the term exclusively applied in France to zigzag lightning. (*Do.*)

(4.) Arago divides lightnings into zigzag, sheet, and fireballs or globular lightning. (*Do.*)

(5.) Arago mentions zigzag lightning generally under the name of "fulminating matter." (*Do. passim.*)

(6.) Lightning is "simply incandescent matter. It indicates the path of the discharge and nothing more." (*Preece*, 338.)

(7.) "Maffei, Chappe, and others, deem that lightning or thunderbolts are almost always elaborated on the ground; that it is from the ground they suddenly dart; that instead of descending from the clouds to the earth, their course is on the contrary from the earth to the clouds. Those who are partisans of this opinion say they have distinctly seen lightning rise like rockets." (*Ar.* 101.)

(8.) "It is difficult to conceive the possibility of distinguishing by the eye whether a flash of lightning between the clouds and the earth rises or descends." (*Do.*)

(9.) There are numerous instances of lightning moving from below upwards. (*Kaem.* 347.)

(10.) The spark probably leaves both bodies at once. Kaemtz has actually observed this in the case of two clouds. (*Do.*)

(11.) "The lightning discharge is the electrical discharge which strikes between a thundercloud and the ground. The latter by the induction from the electricity of the cloud becomes charged with contrary electricity, and, when the tendency of the two electricities to combine exceeds the resistance of the air, the spark passes, which is often expressed by saying that a thunderbolt has fallen." (*Gan.* 828.)

(12.) "Lightning in general strikes from above, but ascending lightning is also sometimes observed; probably this is the case when the clouds being negatively, the earth is positively electrified, for all experiments show that at the ordinary pressure the positive fluid passes through the atmosphere more easily than negative electricity." (*Do.* 831.)

(13.) "It is evident, upon ordinary electrical principles, that if two clouds, or one cloud and the earth, be oppositely excited and charged, the spark and the discharge may either pass from the cloud to the earth or from the earth to the cloud, as circumstances to us imperceptible may direct." (*Lunn*, 9.)

I. D 14—24.

(14.) "People always speak of lightning falling, and never of its rising or ascending, which must often be the case." (*Graves, S. T. E.* 413.)

(15.) "Lightning is the joint work of the positive and negative electricity. When the proper conditions obtain between the earth and the cloud for the production of a flash, then both the positive and negative exert their utmost to approach each other, selecting the easiest available channel, such as trees, bell-wires, church steeples, &c." (*Do.*)

(16.) "Lightning is an enormous electric spark passing between two clouds, or from a cloud to the earth." (*Jen.* 105.)

(17.) "In the latter case, the electrified cloud is attracted towards any prominence or good conductor, which becomes electrified by induction, and the spark of lightning passes when the difference of potential is sufficient to overcome the mechanical resistance of the air." (*Do.*)

(18.) "The thunderbolt as a rule falls only during heavy showers, by taking advantage of the semi-communication with the earth offered by the vertical series of rain drops." (*Nouel. Tel.* 1/9/74.)

(19.) Lightning sometimes escapes from the upper surfaces of clouds. (*Ar.* 40.)

(20.) Lightning exists in our climate without thunder, in a clear sky, in the shape of "heat lightnings." (*Do.* 58.)

(21.) Lightning has been known to occur without thunder in a cloudy sky. (*Do.* 59.)

(22.) Fulminating explosions have been produced by thunderstorms without any luminous appearance. (*Do.* 98.)

(23.) Thunderbolts often develope by their action smoke, and generally a sulphurous odour. (*Do.* 62.)

(24.) "Clouds are not perfect conductors, and therefore do not part with all their discharge at once. Hence a single discharge does not deprive a cloud of the whole of its charge. There may be several successive discharges. The electroscope and galvanometer show this." (*Preece*, 334.)

(25.) "The evidences concerning 'ball lightning' are so numerous that it is impossible to deny that it exists." (*Do.* 368.)

(26.) "Fireballs are among the most interesting and mysterious of electrical phenomena." (*Lat. Clark, S. T. E.* 371.)

(27.) The striking distance of lightning varies from 650 feet to 6,500 feet. (*Mann*, 1875, 531.)

(28.) The duration of the flash does not exceed $\frac{1}{1000}$ of a second. (*Do.*)

(29.) The most brilliant and extensive flashes have not a duration equal to the $\frac{1}{1000}$ of a second. (*Ar.* 48.)

(30.) The following lengths of flashes have been observed:—

In Abyssinia, by M. D'Abbadie	3·7 miles.
In Toulouse, by M. Petit	9·6 ,,
In Weimar, by M. Weissenborn	6·2 ,,

(*Do.* 169.)

(31.) Lightning finds a path of least resistance; hence the zigzags, and the length of the flashes, often several miles. (*Hersch.* 133.)

(32.) Lightning always takes what is technically called the "line of least resistance." (*And.* 142.)

(33.) When lightning strikes an object it frequently divides into several distinct forks or channels. (*Lat. Clark, S. T. E.* 371.)

(34.) "In the tropics a heavily charged cloud does not generally discharge its electricity in a single ray as it does in countries of moderate climate; on the contrary, it frequently takes place from a central point in many radial directions at the same time." (*Treu. S. T. E.* 377.)

(b) *The Action of Lightning on Materials.*

(35.) *Cæteris paribus*, the highest points are those which lightning strikes by preference. (*Ar.* 139.)

(36.) "It is only as it enters metallic bodies, or at the moment of quitting them, that lightning produces much damage." (*Do.* 140.)

(37.) Lightning often fuses metals, not by "cold fusion," but by heating them like ordinary fire. (*Do.* 66.)

(38.) It sometimes contracts metallic wires. (*Do.* 75.)

(39.) It sometimes fuses and then vitrifies earthy substances. (*Do.* 78.)

(40.) Lightning can transport bodies. (*Do.* 86.)

(41.) Its action imparts magnetism; and in passing near a compass needle it affects its magnetism, and sometimes inverts its poles. (*Do.* 88, 89.)

(42.) It forms "fulgurites" by striking sand, fusing it, and giving it the form of vitrified hollow tubes. Some of these tubes have been found 30 or 40 feet long. (*Do.* 79.)

(43.) Fulgurites is fused quartz due to lightning striking sand. (*Hersch.* 135.)

(44.) "The shivering of trees into small splinters like a broom is probably owing to the rarefaction of the sap in the longitudinal pores, or capillary pipes, in the substance of the wood, and the blowing up of bricks or stones in a hearth, rending stones out of a foundation, and splitting of walls, are also probably effects sometimes of rarefied moisture in the earth, under the hearth, or in the walls." (*Frank.* 429.)

(45.) Bodies such as air, glass, resins, dry wood, stones, and substances generally which resist the progress of discharge, involve an explosive form of action attended by light, heat, and an enormous expansive force. (*W. O.* 1858, *App. A.* 2.)

(46.) "It is, in fact, this terrible expansive power which we have to dread in cases of buildings struck by lightning, rather than the actual heat attendant on the discharge itself." (*Do.* 6.)

(47.) With metallic substances the expansive action of lightning ceases. (*Do.* 3.)

(48.) The particular faces of a building that are most exposed to be struck are perhaps those most exposed to the prevailing winds of a country during thunderstorms. (*Ar.* 198.)

(49.) Nollet says that *cæteris paribus*, spires covered with slates are more often struck than those built with stone. This is probably on account of the metal nails and wooden laths used in fixing the slates. (*Do.* 199.)

(50.) It is futile to consider low elevation of buildings as an absolute safeguard. (*Do.*)

(*c*) Return Strokes.

(51.) "Many persons who are killed by lightning are killed by the simple shock resulting from the sudden discharge of electricity from their bodies, which had been inductively electrified from the clouds; the lightning passing from cloud to cloud discharges these, and the escape of the electricity from the body previously charged produces the shock." (*Jen.* 363.)

(52.) "Lord Mahon fused metals and produced strong physiological effects by the return stroke." (*Tynd.* 106.)

(53.) "In nature, disastrous effects may be produced by the return stroke. The earth's surface, and animals or men upon it, may be powerfully influenced by one end of an electrified cloud. Discharge may occur at the other end, possibly miles away. The restoration of the electric equilibrium by the return stroke may be so violent as to cause death." (*Do.*)

(54.) "It was the action of the return shock upon a dead frog's limbs, observed in the laboratory of Professor Galvani, that led to Galvani's experiments on animal electricity, and led further to the discovery by Volta of the electricity which bears his name." (*Do.*)

(55.) Return shock is "a violent and sometimes fatal shock which men and animals experience, even when at a great distance from the place where the lightning discharge passes. This is caused by the inductive action which the thundercloud exerts on bodies placed within the sphere of its activity." (*Gan.* 832.)

(56.) "These bodies are then charged with the opposite electricity to that of the cloud; but when the latter is discharged by the recombination of its electricity with that of the ground the induction ceases, and the bodies reverting rapidly from the electrical to the neutral state, the concussion in question is produced." (*Do.*)

(57.) "The return shock is always less violent than the direct one; there is no instance of its having produced any inflammation, yet plenty of cases in which it has killed both men and animals." (*Do.*)

(58.) "The shocks experienced by living people on the instant of a discharge of lightning without fatal results are generally 'return shocks.'" (*Mann*, 1875, 539.)

(d) *The Effect of Lightning on Persons.*

(59.) People killed by lightning suffer no pain. (*Tynd.* 99.)

(60.) Persons struck see previously no lightning. (*Ar.* 13.)

(61.) The most frequent result of non-fatal strokes is partial paralysis of the legs or arms. (*Do.* 256.)

(62.) Instances have occurred of the hair being burnt off persons in non-fatal strokes. (*Do.*)

(63.) Moderate strokes frequently appear to rid men and animals of maladies. (*Do.* 258.)

(64.) Moderate strokes have been known to accelerate the growth of trees. (*Do.*)

I. D *e* 65—70.

(65.) "*Lightning, Death from.*—The body in these cases retains its warmth for some time, even after death has actually taken place. The first thing to be done is to strip off the clothes and dash cold water upon the trunk in considerable quantities for ten or fifteen minutes; this, in slight cases, will speedily bring about a reaction. The body should also be assiduously rubbed, and artificial respiration be effected. Stimulants of the most active kind are to be resorted to, but that of electricity or galvanism is the one specially called for in this modification of asphyxia." (*Shaw*, 77.)

(*e*) *The Effect of Lightning on Telegraphs.*

(66.) The accidents that telegraphs suffer from might be divided into three classes:—

"1. Those affecting the wires."
"2. Those affecting the poles."
"3. Those affecting the instruments."
(*Preece*, 342.)

(67.) "Each of these classes may be divided into those which are the result of direct discharge, and those which are the result of induction." (*Do.*)

(68.) "In the first case, the wires, poles, or instruments form a path or circuit for a portion of the discharge itself; in the second case, they are influenced by currents which are induced by the approach, recession, or sudden neutralisation of charged clouds." (*Do.*)

(69.) "The direct effects are not nearly so numerous as the induced." (*Do.* 343.)

(70.) As regards induced effects during a thunderstorm, wires are pervaded by repeated currents which ring bells, demagnetise needles, throw apparatus out of adjustment, shock clerks, and make false signals on railway block instruments." (*Do.*)

(71.) During the construction of the Husac tunnel, two premature explosions of blasting charges were caused by induced currents, producing fatal effects. (*Do.*)

(72.) "Not only are the overground wires affected, but those buried two feet deep underground." (*Do.*)

(73.) "Sometimes poles have been shivered, but usually a discharge which has taken the wire in its path divides itself among several poles, and cuts out with the smoothness of a gouge grooves from the top to the bottom." (*Do.*)

(74.) Mr. S. A. Varley's carbon, or "lightning bridge" protector is in use with post-office instruments. (*Do.* 350.)

(75.) In India, where thunderstorms are far more intense than in England, "every office is protected by earth-plates separated from line-plates by a thin layer of air, the interval being maintained by thin ebonite washers." (Dr. Werner Siemens' plate dischargers.) The posts are of iron. Accidents are rare. (*Do.* 352.)

(76.) Some submarine mines used in the defence of Venice and Pola were exploded from the sudden development of induced currents in the mines. (*Abel, S. T. E.* 374.)

(77.) Instruments are sometimes damaged in India by currents not sufficient to charge the upper plate of the Siemens' discharger to a "sparkling potential." (*Ayrt. S. T. E.* 363.)

(78.) An atmospheric discharge of very high potential "will jump from the line wire to the insulator stalks if they be connected with the earth, smashing the insulators in its path." (*Do.*)

(79.) On the 4th July, 1871, at the underground telegraph line on the Manchester and Liverpool railway, whilst a jointer, during a thunderstorm, was jointing some wires, he found that sparks were given off to his jointer's box while resting on the ground. (*Mr. G. E. Preece, S. T. E.* 356.)

(E.) THE INFLUENCE OF METALS ON LIGHTNING.

(1.) One of the objects of Sir. W. Snow Harris's treatise, as stated in his preface, was "to remove the misapprehension which exists as to the attractive effect of metallic bodies in storms of lightning." (*Harr. Pref.*)

(2.) "In every instance of damage to buildings or shipping by lightning, the cause of the electrical discharge is determined in a similar way, through points presenting the least resistance to its progress, and the mischief invariably occurs between detached masses of metal." (*Do.* 88.)

(3.) "When a charged thundercloud hangs low over the earth, all projecting bodies rising from the ground are most powerfully influenced by the inductive aciton, and conducting bodies such as rods, tubes, and sheets of metal more powerfully than bodies that have less conducting capacity." (*Mann*, 1875, 531.)

(4.) Metals have no attraction or affinity for lightning. (*W. O.* 1858, *App.* A. 9.)

(5.) Lightning falls on trees, rocks, and buildings, whether they have metal about them or not. (*Do.*)

(6.) Alluding to the village church of Rosenberg in Carinthia, Austria, which was repeatedly struck in the seventeenth and eighteenth centuries, before rods were known, "very possibly there were large pieces of metal on the wall or in the roof; or if not, there may have been masses of water near, underground, sufficient to account for the manifestation of the electric force." (*And.* 64.)

(7.) "As the use of metals, especially iron, in the construction of buildings, both exterior and interior, is rapidly extending, this becomes a very important consideration in

I. E 8—12; F 1—3.

planning the design of lightning conductors." (*Do.* 140.)

(8.) A human body may form a path of least resistance between insulated metals in a building, and thus may receive a discharge. (*Do.* 160.)

(9.) "Practical experience has pretty fully proved that an electric discharge has a very great affinity for all metal bodies, whether they are insulated from the earth or connected with it." (*Do.* 168.)

(10.) Quoting Dr. Holtz, of Greisswald, "there can be no doubt whatever that the large increase of the use of metals in the construction and ornamentation of modern houses has led to far greater damage to which they are exposed from lightning." (*Do.* 225.)

(11.) "Lightning seeks out by preference metallic substances, whether external or concealed, which are either at or near the point towards which it falls, or near its subsequent serpentine course." (*Ar.* 139.)

(12.) When there are several metallic masses on a roof separated from each other, "it is difficult or impossible to say which of them will be struck by preference," therefore the safe plan is to unite them all metallically to the conductor. (*Do.* 224.)

(F.) PRESERVATIVES FROM LIGHTNING.

(1.) Pliny relates that the ancients believed that thunderbolts never penetrated beyond 6 feet below the surface of the ground, and that they regarded most caves as asylums. (*Ar.* 188.)

(2.) Suetonius relates that the Emperor Augustus retired into a vault when thunderstorms were foreseen. (*Do.*)

(3.) Kæmpfer relates that the Emperors of Japan have similar ideas. (*Do.* 189.)

I. F 4—14.

(4.) The Romans covered their tents with sealskins as preservatives. The Emperor Augustus always wore one. (*Do.* 190.)

(5.) The Chinese consider that mulberry and peach trees preserve against lightning. (*Do.* 191.)

(6.) Mr. Hugh Maxwell, writing in 1787 of his American experience, says of trees:—

Elm . . .	} Often struck.	Ash	} Rarely.	
Chestnut . .		Beech		
Oak . . .		Birch	} Never.	
Pine . . .		Maple		

(*Do.*)

(7.) M. de Thury has met with instances in his French experience of the following trees being struck:—viz. beech, pine, fir, cherry, acacia, elm, oak, and poplar. (*Do.* 192.)

(8.) Pines and resinous trees are said to be less dangerous than oaks and elms. (*Gan.* 131.)

(9.) Men are often struck in the middle of open plains. (*Ar.* 192.)

(10.) Facts show that under trees the danger is still greater. (*Do.*)

(11.) Dr. Winthrop suggested (and Franklin approved the idea) that one ought to place oneself from 16 to 40 feet from a large tree, or intermediate between two trees at this distance. (*Do.*)

(12.) Henley also approved, and, in the case of only one tree, recommended a distance of five or six yards beyond the extremities of the largest branches. (*Do.*)

(13.) The best thing to do in a storm is to lie flat on the ground and not to mind getting wet. (*Mann*, 1878, 338.)

(14.) There is danger in moving when caught in the open air in a thunderstorm, and the inconvenience of getting wet by remaining stationary should be balanced

I. F. 15—21.

against the chances of lessening the risk of being struck. (*Ar.* 199.)

(15.) The danger of being struck is sensibly increased by metals attached to the person. (*Do.* 194.)

(16.) Persons afraid of lightning have made glass cages as asylums. (*Do.* 193.)

(17.) Franklin recommended the following precautions to be adopted in houses :—

> (*a.*) Avoid the neighbourhood of fireplaces, as lightning often enters by the chimney.
> (*b.*) Avoid also metals, including gildings, mirrors, &c.
> (*c*) The best place, consistently with (*a*) and (*b*), is the middle of the room.
> (*d*) The less the contact with walls and floors, the less the danger. A hammock suspended by silken cords in the middle of a large room is good.
> (*e*) Keep generally near substances which are bad conductors, such as silk, glass, pitch, mattresses, &c. (*Do.* 198.)

(18.) Volta thought large fires would prevent thunderstorms. (*Do.* 212.)

(19.) At Caserna, in Romagna, by the advice of the *curé*, the inhabitants on the approach of thunderstorms used to place heaps of straw and brushwood at about every 50 feet, and set them on fire, and for three years they experienced no thunder nor hail. (*Do.*)

(20.) Sailors believe that the firing of artillery disperses thunderclouds, and agriculturists, in some French districts, also think so; but the bombardments of Rio Janeiro in 1711, and of Dannholm, near Stralsund, in 1806, both events that occurred during thunderstorms, failed to disperse, them and equally powerless is the constant practice of the Artillery School at Vincennes. (*Do.* 214.)

(21.) An idea has prevailed in France that ringing

NOTES ON LIGHTNING. 39

I. G 1—8.

church bells during a thunderstorm was efficacious. This was based on superstition, the bells having been blessed. On the contrary, bell-ringers incur danger from lightning. (*Do.* 219.)

(G.) STATISTICAL AND GEOGRAPHICAL NOTES.

(1.) During the 50 years from 1793 to 1843 more than 253 of H.M. ships are known to have suffered. (*Harr. Preface.*)

(2.) Mainly during the 16 years from 1799 to 1815, 150 cases have happened of British ships being struck, 70 seamen have been killed, 133 wounded, 100 lower masts (with corresponding topmasts and smaller spars) have been destroyed, and damage has been done to the value of £100,000. (*Do.*)

(3.) In the 6 years from 1809 to 1815, 30 British line-of-battle ships, and 15 frigates, were more or less disabled from lightning. (*Do.*)

(4.) In 15 months of the years 1829-30, 3 British liners, 1 frigate, and 1 brig, were more or less disabled. (*Do.*)

(5.) The average annual damage done by lightning in England alone was estimated in Nicholson's Journal of Science (in 1843) as £50,000. (*Do.*)

(6.) Previous to 1872, the average annual number of deaths in England from lightning was 18, and in France, 95. (*Preece*, 336.)

(7.) According to Mr. Symons, the meteorologist, in two storms in June 1872, 200 separate accidents came to his notice, involving 10 persons killed, 33 persons injured, 60 houses struck, of which at least 10 were burnt down, 23 horses and cattle, and 99 sheep, killed. (*Do.*)

(8.) Deaths by lightning in England and Wales in each of the years 1877, 1878. (Extract from Registrar-General's Report.)

LIGHTNING.

Division.	District.	Date of Death.	Sex.	Occupation.	All Ages.
		1877.		TOTAL	12
2	Hartley Wintney	26 Nov.	Male	Agricultural Labourer	1
3	Woodstock	16 Aug.	Male	Agricultural Labourer	1
4	Woodbridge	29 Mar.	Male	Agricultural Labourer	1
6	Atherstone	9 May	Male	Labourer	1
7	Chapel-en-le-Frith	7 July	Male	Agricultural Labourer	1
8	Bolton	5 July	Male	Coal Miner	1
9	Skipton	16 May	Male	Hawker	1
9	Huddersfield	8 Aug.	Male	Joiner	1
9	Sheffield	7 July	Female	Wife of Coal Miner	1
9	Aysgarth	15 Oct.	Male	Agricultural Labourer	1
10	Houghton-le-Spring	16 Aug.	Female	Widow of Coal Miner	1
11	Bangor	10 May	Male	Quarry Labourer	1
		1878.		TOTAL	24
1	Woolwich	29 Aug.	Female	Wife of Gardener	1
2	Croydon	13 Aug.	Male	Labourer	1
2	Dartford	24 Aug.	Female	Wife, Bricklayer's Lab.	1
2	Blean	31 Aug.	Male	Labourer	1
2	East Grinstead	23 July	Male	Labourer	1
2	Abingdon	23 Aug.	Male	Bricklayer	1
4	Braintree	12 Aug.	Male	Agricultural Labourer	1
4	Depwade	12 Aug.	Female	Wife, Agric. Labourer	1
5	Liskeard	3 Aug.	Male	Farmer	1
6	Cheltenham	10 Aug.	Male	Carter	1
6	Newcastle-under-Lyme	24 July	Female	Wife of Chair Mender	1
6	Leek	31 Aug.	Male	Labourer	2
6	Atherstone	6 Aug.	Male	Miner	1
7	Bourn	30 Aug.	Male	Son, Agric. Labourer	1
8	Runcorn	6 Aug.	Male	Gardener	1
8	Northwich	24 July	Male	Agricultural Labourer	1
8	Ormskirk	27 June	Male	Son of Farmer	1
8	Burnley	7 Aug.	Male	Son of Labourer	1
9	Skirlaugh	14 May	Male	Son of Farmer	1
9	Scarborough	30 Aug.	Male	Accountant	1
9	Thirsk	16 Aug.	Male	Son of Farmer	1
10	Darlington	20 Mar.	Male	Shoemaker	1
11	Cardigan	20 July	Male	Farmer	1

I. G 9, 10.

(9.) Deaths by lightning registered in Ireland in each of the years 1877, 1878. (Extract from Registrar-General's Report.)

Division.	Superintendent Registrar's District.	Date of Death.	Sex.	Occupation.	All Ages.
		1877.		TOTAL	7
II.	Strabane . . .	21 June	Male .	Labourer	1
II.	Millford . . .	5 June	Female	Farmer's Widow . .	1
III.	Bailieborough .	20 June	Female	Labourer's Daughter .	1
V.	Tullamore . .	20 June	Male .	Tobacco Labourer . .	1
VI.	Swineford . .	19 June	Male .	Son of Landholder . .	1
VI.	Swineford . .	19 June	Female	Daughter of Landholder	1
VIII.	Tralee . . .	14 Sept.	Female	Child of Labourer . .	1
		1878.		TOTAL	6
II.	Manorhamilton	13 April	Male .	Farmer's Son . . .	1
III.	Ardee . . .	16 April	Male .	Farmer's Son . . .	1
VI.	Oughterard . .	23 July	Male .	Landholder	1
VII.	Dungarvan . .	27 June	Male .	Farmer	1
VIII.	Dunmanway .	21 July	Male .	Labourer	1
VIII.	Rathkeale . .	23 July	Male .	Farmer's Son . . .	1

(*Reg. Ire.*)

(10.) Deaths by lightning in England and Wales in the 8 years, 1871—78. (Extract from Registrar-General's Report.)

	1871.	1872.	1873.	1874.	1875.	1876.	1877.	1878.	Total in the eight Years, 1871-78.
ENGLAND .	28	46	21	25	17	19	12	24	192
Division I.	..	1	1	4	1	..	..	1	8
,, II.	5	2	3	3	1	2	1	5	22
,, III.	6	9	1	2	1	1	1	..	21
,, IV.	1	3	..	1	1	2	1	2	11
,, V.	1	2	..	..	2	5	..	1	11
,, VI.	2	5	4	3	3	..	1	5	23
,, VII.	6	5	2	4	1	1	1	1	21
,, VIII.	4	3	4	3	2	5	1	4	26
,, IX.	2	7	2	3	2	..	4	3	23
,, X.	1	5	3	2	2	..	1	1	15
,, XI.	..	4	1	..	1	3	1	1	11

(*Reg. Eng.*)

I. G 11.

(11.) Table showing geographical distribution of fatal thunderbolts in England and Wales, 1861—1878.

No. of Registration Division.	Counties included.	Deaths in the 8 years, 1871—78.	Annual Rate in 8 years, 1871—78, to ten millions living.	Annual Rate in 10 years, 1861—70, to ten millions living.	Order of precedence quâ visitation by fatal thunderbolts.	
					According to the actual No. of cases in the 8 years, 1871—78.	In proportion to the population in the 18 years, 1861—78.
VII.	Leicester, Rutland, Lincoln, Nottingham, and Derby	21	18·05	19·29	5th	1st
III.	Middlesex (extra Metropolitan), Hertford, Buckingham, Oxford, Northampton, Huntingdon, Bedford, and Cambridge	21	17·45	8·03	5th	2nd
IV.	Essex, Suffolk, and Norfolk	11	11·01	13·55	8th	3rd
IX.	Yorkshire	23	11·25	9·97	2nd	4th
X.	Durham, Northumberland, Cumberland, and Westmoreland	15	12·26	5·46	7th	5th
VI.	Gloucester, Hereford, Shropshire, Stafford, Worcester, and Warwick	23	10·15	7·35	2nd	6th
II.	Surrey (extra Metropolitan), Kent (extra Metropolitan), Sussex, Hants, and Berks	22	11·94	4·48	4th	7th
XI.	Wales and Monmouth	11	9·34	6·62	8th	8th
VIII.	Cheshire and Lancashire	26	9·09	4·11	1st	9th
V.	Wilts, Dorset, Devon, Cornwall, and Somerset	11	7·24	3·23	8th	10th
I.	London (including Metropolitan District)	8	2·92	·33	11th	11th
	For England and Wales	192	10·09	6·50	..	..

(12.) In the Government report of 1852, the average number of persons in France stated to be annually killed by lightning is 69. This is probably below the truth. (*Ar.* 134.)

(13.) The following is an imperfect list, culled by Arago from newspapers, of persons killed in a small portion of France:—

Year.	Months.	Numbers of persons killed.	Remarks.
1841	May—October	12	1 on banks of Seine.
1842	May—September	15	4 in a boat at Marseilles, 3 under trees, 1 in a bed.
1843	April—September	16	7 under trees, 3 under a corn-rick.
1844	March—October	21	1 under tree, 2 ringing bells.
1845	May—October	11	3 under trees, 1 ringing bells.
1846	May—September	18	3 under trees, 1 ringing bells.
1848	July—August	4	
1849	March—May	8	2 under trees.

(*Do.* 135.)

(14.) In France, in the 17 years from 1835 to 1852, 1,308 persons were killed. (*Mann*, 1875, 540.)

(15.) In the United States, in 1797, from June to August inclusive, 24 persons were struck, of whom 17 were killed. (*Kaem.* 351.)

(16.) In Prussia, 1,004 persons were killed during the 9 years from 1869 to 1877. (*And.* 170.)

(17.) In Austria, during the 8 years from 1870 to 1877, 1,702 fires were occasioned by lightning. (*Do.* 174.)

(18.) In Switzerland, 33 deaths occurred from lightning in the two years 1876—77. (*Do.* 175.)

I. G 19—30.

(19.) In Sweden, during the 60 years from 1816 to 1877, 664 persons were killed, of whom 15 were in towns, and 649 in the country. (*Do.* 172.)

(20.) In Russia (except Poland and Finland) in the 5 years, 1870—74, 2,270 persons were killed by lightning, of whom 109 were in towns, and 2,161 in the country. (*Do.* 171.)

(21.) During these 5 years, the following fires were occasioned in Russia (except Poland and Finland) by lightning, viz. in towns 93, in the country 4,099. (*Do.* 171.)

(22.) The year 1880 appears to have been one of general terrestrial electrical disturbance in Europe; thunderstorms were frequent; earthquakes and volcanic eruptions occurred; and auroræ, earth currents, and waterspouts were manifested. Earthquakes also occurred during this year in Smyrna, in the Philippine Isles, and in Chili.

(23.) In Lima, and in Lower Peru generally, there are no clouds, and no thunder and lightning, but a permanent opaque vaporous fog. (*Ar.* 109.)

(24.) No rain falls in lower regions of Peru, and a S.W. wind prevails. (*Enc. Met.* 111.)

(25.) Beyond 75° lat. thunderstorms are unknown in the open sea. (*Ar.* 110.)

(26.) In Iceland, they are very little known. (*Do.* 111.)

(27.) In St. Helena, they are very little known. (*Do.* 123.)

(28.) It appears that there are frequently slight shocks of earthquake in the valley of the Cabul river, in the Punjab and Afghanistan; but never thunderstorms.

(29.) In Jamaica, from November to April, lightning and thunder are of almost daily occurrence at the summits of the Port Royal Mountains. (*Ar.* 115.)

(30.) In Devonshire, Cornwall, and the neighbourhood of Swansea, thunderstorms are frequent in proportion to the absence of metallic mines, a fact probably due to their furnaces and tall chimneys. (*Do.* 116.)

I. G 31—39.

(31.) In America there is an opinion that barns full of grain or forage are more often struck than other buildings. (*Do.* 199.)

(32.) Thunderstorms disperse, or turn off, at Niort, Mayenne, France, where much diorite containing iron exists. (*Do.* 117.)

(33.) At Grondôme, in the Apennines, there is a pointed eminence containing iron in serpentine rock, where frequent short thunderstorms occur. (*Do.*)

(34.) In New Granada, the position of Tumba Barreto, near the gold mine of Vega de Sapia, is avoided by the miners, many having been killed there by the frequent strokes of lightning. (*Do.* 180.)

(35.) "In the interior of the great towns of Europe, men appear to be very little exposed to danger from lightning." *Do.* 178.)

(36.) "According to an opinion widely prevailing, persons are much more exposed in villages and in the open country." (*Do.* 180.)

(37.) In the Mediterranean, in 15 months of the years 1829—30, 5 of H.M. ships were struck. (*Ar.* 186.)

(38.) Table of frequency of thunderstorms.

Place.	No. of years observations.	Years in which observed.	Mean No. of days per year.
Paris	41	1785 to 1837	13·6
London	13	1807 „ 1822	8·3
Berlin	15	1770 „ 1785	18·3
St. Petersburg	11	1726 „ 1736	9·1
Toulouse	7	1784 „ 1790	15·4
Padua	4	1780 „ 1783	17·3
Cairo	2	1835 „ 1836	3·5
Rio Janeiro	6	1782 „ 1787	50·6
Pekin	6	1757 „ 1762	5·8

(*Do.* 128.)

(39.) Thunderstorms are more numerous in Schleswig-Holstein than in any other part of Central and Northern

I. G 40—42.

Europe. The province is intersected by rivers and canals. (*And.* 222.)

(40.) On August 9th, 1863, Mr. E. Whymper, in attempting to ascend the Matterhorn, experienced, near its summit, a severe snowstorm which lasted 26 hours, and shortly after its commencement became a heavy thunderstorm, with much thunder and lightning all round his position. All this time fine weather prevailed below the mountain, only a small cloud having been observed near its summit. (*Whymp.* 169.)

(41.) On July 30th, 1869, Mr. R. B. Heathcote, of Chingford, Essex, was with 3 guides within 500 feet of the top of the Matterhorn, when he was surrounded by mist and heard close to him much thunder. There was no wind nor rain at the time, nor apparently did he see any lightning. The Matterhorn, like all Alpine rock summits, is frequently struck by lightning. (*Whymp.* 414.)

(42.) On the 15th July, 1880, at 1.30 a.m., an explosion occurred at the Risca coal pits near Newport, Monmouthshire, "while a tremendous thunderstorm was raging over the district." The shaft is 280 yards deep to the landing place. The men at the time in the mine, numbering 119, were killed, and the engine house, fan, and adjacent shaft were greatly damaged. "Very vivid and frequent lightning was observed." The cause of the explosion has not been ascertained, but it was conjectured at the time that "the electrical condition of the atmosphere above ground may have had something to do with it." (*Stand.* 16/7/80. *Graph.* 24/7/80.)

CHAPTER II.—NOTES ON LIGHTNING ENGINEERING.

(A.) HISTORICAL NOTES.

(1.) "MANY centuries before Christ it had been observed that yellow amber (*elektron*), when rubbed, possessed the power of attracting light bodies. . . . . This is the germ out of which has grown the science of electricity, a name derived from the substance in which this power of attraction was first observed. This attraction was the sum of the world's knowledge of electricity for more than 2,000 years." (*Tynd.* 1.)

(2.) In A.D. 1600, Dr. Gilbert observed that various spars, gems, stones, glasses, and resins, possessed the same power as amber. (*Do.* 2.) [*See also I. C.* 63.]

(3.) In 1675, Robert Boyle observed that rubbed amber became itself attracted. He also saw the light of electricity by rubbing a diamond. (*Do.*)

(4.) About 1675, Otto von Guericke, Burgomaster of Magdeburg, and inventor of the air-pump, devised the electrical machine in the form of a ball of sulphur. He also noticed the power of repulsion. (*Do.*)

(5.) In 1675, electric light was first observed by Pickard, and in 1705, John Bernouilli and Hawksbee experimented on it. (*Do.*)

(6.) In 1708, Dr. Ward experimented with amber and wool, and produced cracklings and light; and he says, "This light and crackling seem in some degree to represent thunder and lightning." This is the first published allusion to thunder and lightning in connection with electricity. (*Do.* 3.)

II. A 7—13.

(7.) In 1729, Stephen Gray also observed the electric spark, and that the power evolved seemed to be of the same nature with that of thunder and lightning. In the same year he first noticed the actions of conduction and insulation. (*Do.* 13 and 14.)

(8.) In 1733—37, the influence of moisture as a conductor was first demonstrated by Du Fay, who sent a charge through 1,256 feet of pack-thread. He also discovered that there are two kinds of electricity, vitreous and resinous. (*Do.* 22.)

(9.) On the 23rd January, 1744, Ludolf, at the opening of the Academy of Sciences at Berlin by Frederick the Great, first ignited substances by the electric spark. (*Do.* 80.)

(10.) On the 4th November, 1745, Kleist, a clergyman of Cammin, in Pomerania, announced by letter to Dr. Lieberkühn, of Berlin, the discovery of the principle of the Leyden jar. Kleist missed the explanation of the principle, but the Leyden philosophers gave it, whence it derived the name of "Leyden jar." (*Do.* 66.)

(11.) In 1748, Dr. Watson and Dr. Bevis improved the apparatus so as to arrive at the form of the present day. (*Do.*)

(12.) Benjamin Franklin was born at Boston in January, 1706. He was a printer by trade, and kept a general store at Philadelphia for some time. He first heard of electricity at a lecture by Dr. Spence in 1746. He then investigated the subject and made experiments himself, sending accounts of them to the Royal Society. In 1750 he reported on the identity of electricity and lightning, and submitted the idea of fixing sharp pointed iron rods to the summits of buildings in order to protect them from lightning. Buffon, the great naturalist, had Franklin's pamphlet on the subject translated, and through this means it spread over Europe. (*And.* 18.)

(13.) At Buffon's instigation, M. D'Alibard made experiments, according to Franklin's suggestions, with a

II. A 14—19.

pointed iron rod 1 inch in diameter and 80 feet long, at Marly, near Paris; and on the 10th May, 1752, during a thunderstorm, the custodian of the rod (during D'Alibard's absence) observed at it vivid flashes. (*Do.* 20.)

(14.) Unconscious of D'Alibard's success, Franklin himself experimented with a kite having a thin iron wire 1 foot long at the top, and on the evening of the 4th July, 1752, during a thunderstorm, he drew flashes with this kite in a field near his house at Philadelphia. (*Do.* 23.)

(15.) In 1752, Franklin erected an iron lightning rod at his house, with a sharp steel point projecting 7 or 8 feet above the roof, and with the other end 5 feet in the ground By means of a contrivance, two bells were rung whenever an electrical current passed through the rod. He found that the bells rang sometimes "when there was no lightning or thunder, but only a dark cloud over the rod;" sometimes, after a flash of lightning they would suddenly stop; and at other times, when they had not rung before, they would suddenly begin to ring; and there were considerable fluctuations in the currents. (*Do.* 26.)

(16.) Franklin advertised, recommending people to erect their own iron rods, and giving simple directions; houses thus armed were, however, found to be struck by lightning owing to the rods having been put up wrongly, or by impostor professors. (*Do.* 29.)

(17.) "The untaught multitude and the bigoted zealots opposed in Europe, as they did in America, the establishment of lightning conductors." To these was added "a not numerous but powerful section of literary men," chiefly French. The Abbé Nollet (teacher in Natural Philosophy) considerably retarded the introduction of lightning rods. (*Do.* 35 and 18.)

(18.) Mr. Wilson, against the opinion of Franklin, Cavendish, and Watson, advocated the use of blunt conductors. (*Tynd.* 101.)

(19.) Franklin devised the theory of a single electric

II. A 20—28.

fluid to explain electrical phenomena. Symmer devised that of two fluids, which is simpler. (*Do.* 22.)

(20.) The inscription on a medal dedicated to Franklin after the declaration of American Independence was *Eripuit fulmen cœlo, sceptrumque tyrannis.* (*Hersch.*)

(21.) In 1753, Professor Richmann was killed at St. Petersburg by "a thundercloud which discharged itself against the external rod" of his apparatus. (*Tynd.* 101.) [*See* III. 38.]

(22.) In 1757, De Romas, at Nérac, in France, sent up a kite 400 or 500 feet into the air during a moderate storm, and obtained thirty flashes, 9 or 10 feet long and 1 inch broad, in less than an hour, besides a thousand of 7 feet long and under. (*Ar.* 234 and *Hersch.*)

(23.) About 1772, Beccaria experimented, during thunderstorms, at Valentino Palace, Turin, with rods whose lower portions, connected with the ground, were slightly separated from the pointed upper parts, and obtained vivid sparks across the gap. (*Ar.* 230.)

(24.) The destruction of the steeple of St. Bride's Church in London on the 18th June, 1764, led Dr. Watson to force the claims of rods on public attention in England. He erected the first rod in England at his own cottage at Payneshill, near London, in 1762. (*And.* 38.) [*See* III. 62.]

(25.) The first lightning rod placed on a public building in Europe was at St. Jacob's Church, Hamburg, in 1769. (*Do.* 43.)

(26.) In 1769, St. Paul's Cathedral was protected by lightning rods. (*Tynd.* 102.) [*See* III. 145.]

(27.) The fact of Purfleet storehouse, when defended by a rod, being struck and damaged on the 15th May, 1777, led to an outcry against pointed rods. (*And.* 41.) [*See* III. 39.]

(28.) In 1779, Lord Mahon (afterwards Lord Stanhope) published his "Principles of Electricity," with an explanation of "returning strokes." (*Tynd.* 102.)

II. A 29—35.

(29.) In 1829, solid iron rods 1½ in. diameter, with tops of copper tipped with gold, were ordered to be used for the War Department powder magazines. (*A. M., W. O.* 1829.)

(30.) "The most decisive evidence in favour of conductors was obtained from ships." Sir William Snow Harris's fixed lightning rods for ships were of great value. (*Tynd.* 102.)

(31.) In 1820, Sir W. Snow Harris turned his attention to the best means of protecting H.M. ships from damage by lightning. In 1839, a Government Commission was appointed, and on their report Harris's system was adopted. In 1855, he designed a system of protection for the new Houses of Parliament at a cost of £2,314, consisting mainly of copper tubes and bands. In 1858, he was called upon by the War Office to give advice as regards protecting powder magazines, and his recommendations were adopted. (*And.* 85—98.)

(32.) In 1822, the French Minister of the Interior ordered all public buildings to be protected by rods, and applied to the Academy for advice; this body nominated a Committee, who, through M. Gay Lussac, one of their number, presented a report, dated 23rd April, 1823, which laid down rules for guidance as to the area over which rods possessed a protective power. (*Do.* 76.)

(33.) "When thirty years had passed, the instances of buildings "armed with conductors" being struck became so numerous that it was impossible to ignore them any longer, and another Committee of the Academy was selected; their report was made on the 18th December, 1854, and was drawn up by Professor Pouillet. (*Do.* 78.)

(34.) On account of the Palace of the Louvre having been struck, though armed with rods, a third Committee of the French Academy assembled, and presented (again through Professor Pouillet) a report dated 19th February, 1855. (*Do.* 81.)

(35.) Twelve years afterwards, several powder magazines with rods on them having been struck, a fourth

II. B 1—8.

Committee of the Academy was nominated. Marshal Vaillant, Minister of War, and M. Becquerel, were members of it. Professor Pouillet was again the author of their report, which was dated 14th January, 1867. (*Do.* 82.)

(B.) DETAILS OF LIGHTNING RODS.

(1.) The disadvantages of rigid bars are now (1855) avoided by the use of flexible metal cords. (*Ar.* 247.)

(2.) According to facts collected, square or cylindrical bars of iron $\frac{8}{10}$ inch in diameter are sufficient to prevent fusion. (*Do.*)

(3.) Abrupt changes of course in rods should be avoided. (*Do.* 250.)

(4.) A rod of copper $\frac{1}{2}$ inch in diameter, and 6 inches long, has never been fairly melted; and much less dimensions have carried away heavy discharges. (*Harr.* 113.)

(5.) Harris's practical deductions regarding rods were:

(*a.*) Copper was the best metal to use.

(*b.*) It should be not less than $\frac{1}{2}$ inch in diameter.

(*c.*) The principal detached metallic masses of the building should be connected to it.

(*d.*) The rods should be attached to the most prominent points.

(*e.*) They should be carried close to the walls and directly into the ground. (*Do.* 123.)

(6.) Ships' conductors were at first iron chains, or long links of small iron rods, carried from the head of the mast over the side of the ship into the sea. (*Do.* 130.)

(7.) In 1762, at Dr. Watson's suggestions, H.M. ships were supplied with rods formed of long links of copper rod $\frac{1}{4}$ inch in diameter, united by small eyes, and carried from the top-gallant mast head into the sea as before. (*Do.* 131.)

(8.) It was found that linked and chain rods were bad,

II. B 9—18.

as the links were often broken and the metal fused. (*Do.* 133.)

(9.) In 1824, the French used wire rope conductors for their ships. (*Do.* 134.)

(10.) All these conductors, being parts of the rigging, were found to be dangerous. Seamen came unawares into their bights when the top-gallant masts were struck, &c., and thus formed portions of the lines of least resistance for the discharge. Besides, being movable and generally stowed away till thunder weather approached, the rods were not always found when wanted, nor applied in time, nor properly. (*Do.* 137.)

(11.) In 1821, Sir W. Snow Harris proposed to incorporate the rods in the masts of the ship, thus practically making them fixed, and to connect them with the metals of the hull, and finally with the sea by means of the keel and kelson bolts. (*Do.* 141.)

(12.) His rods were bands of copper $\frac{1}{16}$ to $\frac{1}{8}$ inch thick, and $1\frac{1}{2}$ to 5 inches wide, let into the masts. (*Do.* 142.)

(13.) This system was, about 1831, adopted into the British Navy, and it proved very successful in reducing the number of accidents to ships from lightning. (*Do.* 148.)

(14.) A similar system was proposed for merchant ships, except that from the heads of the lower masts, wire ropes, $\frac{1}{2}$ inch in diameter, were let down the rigging, and over the ship's side into the sea. (*Do.* 151.)

(15.) The French employ for buildings copper ropes $\frac{4}{10}$ to $\frac{8}{10}$ of an inch in diameter for each 82 feet in height. (*Mann*, 1875, 534.)

(16.) "Conductors require to be of larger size in proportion to their length." (*Do.*)

(17.) Dr. Mann recommends "a rope of galvanised iron, consisting of 42 strands of $\frac{1}{10}$-inch wire with a brush point." (*Do.*)

(18.) "Conductors require to be expanded and amplified both at their summits and at their roots or base." (*Do.* 533.)

II. B 19—29.

(19.) The French electricians recommend a cluster of points instead of one alone. (*Do.* 536.)

(20.) "What the dimensions in a lightning conductor are that would fulfil this essential condition of giving sufficient capacity for the safe transmission of the largest possible discharge, is yet an unsettled question." (*Do.* 533.)

(21.) M. Melsen aims at covering a building with a sort of metallic net, with numerous points and earth contacts. At the Hôtel de Ville at Brussels his system has been adopted. This building has a pinnacled spire 297 feet high. The main conductors are 8 galvanised iron rods, each about $\frac{2}{8}$ inch in diameter and 310 feet long. These in their course gather up strands of similar size from the ridges, parapets, turrets, and gables. (*Do.*)

(22.) M. Melsen's experiments tend to show that copper is less strong than iron to resist the disintegrating effects of a powerful discharge. (*Do.* 1878, 334.)

(23.) A copper rope should not be less than $\frac{4}{10}$ inch, and a galvanised iron one not less than $\frac{8}{10}$ inch in diameter. (*Do.* 335.)

(24.) Conductors should have no joints except well-soldered ones. (*Do.* 346.)

(25.) Chains and linked rods should be strenuously avoided. (*Do.*)

(26.) The straighter conductors are the better, lest branch circuits should be determined. (*Do.*)

(27.) "A wire equivalent to the ordinary galvanised iron wire known as No. 4, which is $\frac{1}{4}$ inch in diameter, and used so largely for telegraphic purposes, is amply sufficient for any dwelling-house." (*Preece*, 344.)

(28.) The above opinion is based on experience, derived from the immunity from injury by lightning obtained by telegraph poles supplied with No. 8 earth wires (which are only half as massive), whose upper extremities terminate in points. (*Do.*)

(29.) Mr. Preece "can conceive no case in which $\frac{1}{4}$ inch stranded galvanised iron wire is not ample." (*Do.* 346.)

(30.) Mr. Latimer Clark agreed with Mr. Preece as to the use of No. 4 galvanised iron wire for ordinary houses where economy had to be studied. (*Lat. Clark, S. T. E.* 373.)

(31.) Copper is the best material for rods. (*W. O.* 1875, 11.)

(32.) Iron may be employed, but larger dimensions must be used, so as to give the same conductivity as copper. (*Do.* 12.)

(33.) Copper conductors should be as follows :—rods $\frac{1}{2}$ inch diameter, tubes $\frac{5}{8}$ inch diameter and $\frac{1}{8}$ inch thick, bands $1\frac{1}{4}$ inches wide and $\frac{1}{8}$ inch thick. (*Do.* 13.)

(34.) Iron conductors should be solid rods 1 inch in diameter, or bands 2 inches wide and $\frac{3}{8}$ inch thick. (*Do.* 14.)

(35.) Copper is, as regards iron, more conductive, less corrodible, more expensive, more liable to mechanical injury, more liable to be stolen, and more fusible. (*Do.* 15.)

(36.) Roughly speaking, copper and iron cost the same for the same conductivity. (*Do.* 16.)

(37.) The expansion and contraction of the metal must be provided for, especially at the joints. (*Do.* 17.)

(38.) Where contact of two different metals occurs, precautions should be taken to prevent access of moisture, and hence galvanic action and decomposition. (*Do.* 25.)

(39.) Ample surface is essential to rods in order to give the discharge free room for expansion. (*Do.* 1858, *App. B.* 1.)

(40.) Details of conductors affixed to Weedon powder magazines in 1857, under the personal direction of Sir W. Snow Harris, are given as aids in forming specifications and estimates, viz. :—

Solid copper pointed rods, $\frac{1}{2}$ inch diameter, 5 feet high,
Flat ridge conductors, 4 inches by $\frac{1}{8}$ inch.
Tubular gable conductors, 1 inch by $\frac{1}{8}$ inch.
Flat vertical conductor, 3 inches by $\frac{1}{8}$ inch.

II. B 41—46.

Copper rain-water down-pipes.

Underground copper earth connections, 3 inches by $\frac{1}{8}$ inch, and 4 feet long to forks or branches.

Copper forks of earth connections, each 2 inches by $\frac{1}{8}$ inch and 4 feet long.

Flat copper bands, from copper sheathing on doors to rain-water pipes, 2 inches by $\frac{1}{8}$ inch. (*W. O.* 1858, *and R. E. A.* 89.)

(41.) As a general rule all the joints should be covered with a layer of solder at least 1 inch thick. (*Franc. Mich. Tel.* 64.)

(42.) "Copper, much less resisting than iron in a mechanical point of view, undergoes with rapidity, under the influence of electric currents and atmospheric variations, both a kind of disaggregation and temper, rendering it fragile and brittle, *i.e.* in a very short time its primitive solidity is much altered." (*Do.*)

(43.) There are two principal parts in a lightning rod, the terminal rod and the stalk; the terminal rod is a pointed bar of iron from 6 to 10 feet high, fixed vertically to the roof of the edifice to be protected; and its basal section is about 2 or 3 inches in diameter. (*Gan.* 833.)

(44.) The stalk is best formed of wire cord of half a square inch in section, as being less rigid. Copper wire cord is recommended by the French Academy of Sciences. (*Do.*)

(45.) The conditions for lightning rods are:—

- (*a*) So large as not to be melted.
- (*b*) To terminate in a point.
- (*c*) To be continuous, and to have intimate connection with the ground.
- (*d*) The metallic surfaces of the building to be connected to it. (*Do.*)

(46.) "If the last two conditions are not fulfilled there

is great danger of lateral discharges; that is to say, that the discharge takes place between the conductor and the edifice, and then it only increases the danger." (*Do.* 834.)

(47.) "The first conductors were invariably rods of iron, this metal being preferred by Franklin and his immediate followers as cheap, ready at hand, and answering all purposes in practice." (*And.* 50.)

(48.) The French Committee of the 18th December, 1854, recommended greater capacity for rods and as few joints as possible, and all joints to be tin soldered. (*Do.* 79.)

(49.) Brass is not so good a conductor as copper; it is also very liable to destruction by atmospheric influences; and it was discarded altogether some thirty or forty years ago. (*Do.* 107.)

(50.) The cost of copper is seldom less than six or seven times that of iron. (*Do.* 108.)

(51.) Till recently it was very difficult to manufacture long rods or bands of pure copper. The Spanish "Rio Tinto" copper is barely equal to iron in conductivity. The difficulty of obtaining pure copper was solved by the demand for submarine cables. (*Do.* 110.)

(52.) The lightning rods made at the works of Mr. R. S. Newall, F.R.S., at Gateshead-on-Tyne, established forty years ago, have generally a conductivity of 93 per cent. of pure copper. (*Do.*)

(53.) M. Melsen prefers iron to copper, since it has more molecular strength and would resist a great charge better. (*Do.* 117.)

(54.) The Victoria and Clock Towers of the Houses of Parliament, each three hundred feet high, have copper bands 5 inches wide and ¼ inch thick, connected with the roof metal, and with a metallic staircase in each tower. The roofs of Westminster Palace are covered with galvanised iron, and, in many cases, connected to the earth by cast-iron water-pipes. (*Do.* 119.)

(55.) In France lengths of iron bars are principally used as lightning rods. The whole rod is covered with tar or

II. B 56—60.

paint, except the terminal, in order to preserve it from the air. Sometimes galvanised iron cables of 1 inch diameter, or red copper cables of ½ inch diameter are used. (*Do.* 129 and 132.)

(56.) In France "all masses of metal used in the construction of the building are metallically connected with the *paratonnerres*. As a rule this is done by pieces of iron about ½ inch square, which are strongly soldered to the metal surfaces, and then connected with some part of the conductor or ridge circuit." (*Do.* 130.)

(57.) In America the conductors are of iron. (*Do.* 134.)

(58.) "After repeated experiments, Mr. R. S. Newall has arrived at the conclusion that a conductor made of copper of adequate size is the best, and, in the end, the cheapest means of protecting buildings from the effects of lightning." (*Do.* 143.)

(59.) "For private houses and buildings, a rope made of copper ought to be at least ⅝ inch in diameter. For chimneys of manufactories where gases are liable to corrode the rope, it had better be a little thicker." (*Do.* 151.)

(60.) In June and July, 1880, Mr. W. H. Preece, with Dr. Warren de la Rue's battery of 3,240 chloride of silver cells (whose charge was accumulated in a condenser of a capacity of 42·8 microfarads, whereby a potential of 3,317 volts was obtained), transmitted currents through copper conductors (and afterwards through leaden ones) of precisely the same mass, but in three different forms, viz. those of a solid cylinder, a tube, and a ribbon. The currents passed through platinum wire of ·0125 inch diameter, of various lengths, which were calculated to display by the character of their deflagration, or by the shades of heat manifested, any difference (not less than 5 per cent.) that might have existed in the strength of the discharges due to the difference in the forms of the conductors. No difference was apparent in any of the currents; and the conclusion arrived at by Mr. Preece was "that the discharges of electricity of high potentials obey the law of Ohm, and are not affected by change of form. Hence, extent of surface does not favour lightning discharge. No more efficient lightning conductor than a cylindrical rod or a wire rope can therefore be devised." (*Tel.* 1*st September*, 1880, *Paper read by Mr. Preece before British Association.*)

(C.) POINTS OF RODS.

(1.) Franklin only required the tops of rods to rise a little above the tops of the chimneys. (*Ar.* 240.)

(2.) Cavendish, Priestley, and other English physicists fixed the height of the rod above the house at 10 feet. (*Do.*)

(3.) "In France (1855) our builders go up to 10 mètres (32 feet 6 inches), and even only stop there on account of considerations connected with solidity." (*Do.*)

(4.) In 1790 the Philosophical Society of Philadelphia approved of Mr. R. Patterson's plan of making points of plumbago. (*Do.* 244.)

(5.) In France (1855) a single point is always used. (*Do.*)

(6.) In Germany and England (1855) sometimes a single point, and sometimes a brush of points, is used. (*Do.*)

(7.) One reason alleged for the brush point was, that if all the points became blunt, their combined action would still equal that of a good single point. (*Do.*)

(8.) Another was, that on account of their divergency, one or other of them would always present itself more or less directly to an approaching thunderstorm. (*Do.*)

(9.) Arago preferred himself the single point, as recommended by Franklin. (*Do.* 245.)

(10.) Points were gilded because, being of iron, they soon became rusted; but this not being found durable, copper gilt points were screwed on; and lastly, platinum points were used. (*Do.* 243.)

(11.) Discharges "occasionally become so modified by

II. C 12—19.

various circumstances as to assume a mere progressive and quiet form often free from any attendant danger whatever. If a pointed metallic rod project from one of the terminating planes of a charged system into the intervening dielectric, the consequence frequently is a luminous brush of beautifully coloured light. The discharge is progressive and occupies a sensible time." (*Harr.* 18.)

(12.) "Any termination which can be conveniently given to a conductor, if even it were a ball of 1 foot in diameter, would be, in relation to perhaps 1,000 acres of cloud, virtually a pointed conductor." (*Do.* 117.)

(13.) Glass balls were placed on the summits of the conductors of some of the lighthouses in lieu of points, it being considered that glass would repel the electric discharge. (*Do.* 129.)

(14.) Franklin experimented on the power of points, and inferred that pointed rods robbed thunderclouds of their charge, and made them shrink back. (*Do.* 187.)

(15.) "The pointed termination of the conductor is a matter of some practical importance, because it establishes a slow and gentle discharge of an accumulation of electrical force at high tension." (*Mann*, 1875, 535.)

(16.) In Paris the rods usually project 12 to 30 feet above the buildings. (*Do.*)

(17.) Professor Gavarret has shown by experiments on the function of points, that the tension which can be produced in a charged conductor, towards which a point is directed, diminishes according as the point is made more acute. (*Do.*)

(18.) Platinum points are specially made for conductors in Paris. (*Do.*)

(19.) At the Hôtel de Ville, Brussels (where the system of protection is designed by M. Melsen), there are "426 points distributed in 60 aigrettes, and lying in 10 different places, provided as air terminals." Of these, 385 are of copper, 63 of galvanised iron, and 8 are spikes gilded at the end. (*Mann*, 1878, 330.)

NOTES ON LIGHTNING ENGINEERING. 61
II. C 20—32.

(20.) Tips of silver alloy are recommended for points. (*Do.* 335.)

(21.) The larger the building, the more points should be used. (*Do.*)

(22.) Terminals should branch out into several points or aigrettes. (*Do.*)

(23.) Points should project into the air at least 8 feet. (*Do.*)

(24.) A projection of 3 feet beyond the chimney-top is sufficient. The point can then be inspected. (*Preece*, 348.)

(25.) The principal function of a conductor depends on its point. (*Do.*)

(26.) As regards the action of points on charge, Faraday made an experiment demonstrating that when a charge passed through the air, this action "simply converted the line of discharge into a conductor; it was precisely the same as if a wire connected the two points." (*Do.* 378.)

(27.) The electric discharge is "simply due to the fact of two points separated by air being raised to such a difference of potential that the resistance of the air cannot restrain their neutralisation across it. Now if we prevent the increase of potential by dissipating the charge as it collects we effect this object. This is the function of points." (*Do.* 353.)

(28.) "Points prevent the accumulation of charge. A pointed body, such as a needle, directed towards a charged conductor dissipates the charge at once. A pointed conductor directed towards a charged cloud dissipates the charge in its immediate neighbourhood quietly and silently." (*Do.*)

(29.) "Thus it is that the chief merit of a lightning conductor consists in the action of its pointed end." (*Do.*)

(30.) "A properly constructed conductor may be said to prevent electric discharge within the sphere of its action." (*Do.*)

(31.) A glow or brush may be seen at the top of a lightning rod if it is in good order. (*Do.*)

(32.) A fork or brush of three or four points may be

II. C 33—39.

used on the tops of rods when the rods are widely separated, or on single prominent points. (*W. O.* 1875, 34.)

(33.) Pointed terminations are so far useful that they "tend to break the force of the discharge of lightning when it falls on them." In fact, before the explosion takes place, a large amount of the discharge which would otherwise take part in the explosion runs off through the points. (*Do.* 1858, *App. A.* 12.)

(34.) Gilding points is of no use. (*Do.*)

(35.) It is sufficient if the terminating rods are roughly pointed. (*Do.*)

(36.) The use of brush points is recommended, since, "radiating in all directions, they will hasten the neutralisation of the electrified cloud; and in the event of a discharge, the discharge, by dividing amongst them, will prevent their fusion." (*Franc. Mich. Tel.* 44.)

(37.) "If you impart a good charge to a sphere, you may figure the electric fluid as a little ocean encompassing the sphere, and of the same depth everywhere. . . . But supposing the conductor to be a cube, an elongated cylinder, a cone, or a disc, the depth, or, as it is sometimes called, the density, of the electricity, will not be everywhere the same. The corners of the cube will impart a stronger charge to your carrier than the sides. The end of the cylinder will impart a stronger charge than its middle. The edge of the disc will impart a stronger charge than its flat surface. The apex or point of the cone will impart a stronger charge than its curved surface or its base." (*Tynd.* 51.)

(38.) Professor Riess, of Berlin, found that he could deduce with great accuracy the sharpness of a point from the charge which it imparted. He compared in this way the sharpness of various thorns with that of a fine English sewing needle. (*Do.* 53.)

(39.) "Considering that each electricity is self-repulsive and that it heaps itself upon a point in the manner here shown, you will have little difficulty in conceiving that when the charge of a conductor carrying a point is sufficiently strong, the electricity will finally disperse itself by streaming from the point." (*Do.*)

(40.) "Flames and glowing embers act like points; they also rapidly discharge electricity." (*Do.* 58.)

(41.) The point is usually of platinum or gilt copper. (*Gan.* 833.)

(42.) "The action of a lightning conductor depending on induction and the power of points, Franklin, as soon as he had established the identity of lightning and electricity, assumed that lightning conductors withdrew electricity from the clouds; the converse is the case." (*Do.*)

(43.) "When a stormcloud positively electrified, for instance, rises in the atmosphere, it acts inductively on the earth, repels the positive, and attracts the negative, fluid, which accumulates in bodies placed on the surface of the soil the more abundantly as these bodies are at a greater height. The tension is then greatest on the highest bodies, which are therefore most exposed to the electrical discharge; but if these bodies are provided with metallic points like the rods of conductors, the negative fluid withdrawn from the soil by the influence of the cloud flows into the atmosphere and neutralises the positive fluid of the cloud." (*Do.*)

(44.) "Hence, not only does a lightning conductor tend to prevent the accumulation of electricity on the surface of the earth, but it also tends to restore the clouds to their natural state, both which concur in preventing lightning discharges." (*Do.*)

(45.) "The disengagement of electricity is, however, sometimes so abundant that the lightning conductor is inadequate to discharge the ground, and the lightning strikes; but the conductor receives the discharge in consequence of its greater conductivity, and the edifice is preserved." (*Do.*)

(46.) The French Committee, who reported through Professor Pouillet on the 18th December, 1854, recommended the points to be rather blunt, as sharp points ran

II. C 47—55.

risks of becoming fused; also that they should be of copper, not of platinum, since the copper became less heated, was a better conductor, and cheaper. (*And.* 79.)

(47.) The point in France is sometimes of platinum, but generally of pure red copper, or an alloy of silver and copper. (*Do.* 127.)

(48.) In America, the terminal rod generally projects about 4 feet above the highest point of the building, and consists of round iron. The upper end is not always pointed. (*Do.* 134.)

(49.) Mr. Newall's terminal rods are from 3 to 5 feet high, and from $\frac{5}{8}$ to $\frac{3}{4}$ inch in diameter. At the top they form a brush of 4 points. (*Do.* 144.)

(50.) The German terminal rod or *Aufgangstange* is of iron, and varies from 10 to 30 feet high, and has a single point. (*Do.* 145.)

(51.) If a prominence on the earth electrified by a cloud "be armed with a point connected with the earth, then, as soon as the potential of the point is raised, even slightly, the electricity passes off from the point into the air; the prominence can no more be highly electrified than a leaky bucket can be filled with water." (*Jen.* 105.)

(52.) "The brush discharges, whether luminous or otherwise, are due to the accumulation of electricity in large quantities at points." (*Do.* 92.)

(53.) "The brushes or sparks which fly off from points charged to high potential, show that in all apparatus intended to remain charged at a high potential every angle and point must be avoided on the external surfaces." (*Do.* 105.)

(54.) "Fix a fine metal point to the conductor of the electric machine and work the machine. It will be impossible to collect any appreciable charge on the conductor; the electricity all escapes by the point." (*Gord. I.* 22.)

(55.) There is a very much greater force tending to drive electricity from a point than from any other portion of a conductor. (*Do.* 23.)

II. D 1—12.

(D.) Earth Connections of Rods.

(1.) Mr. Hare, Professor of Chemistry in Philadelphia, proposed underground iron water-pipes as the earth connections of lightning rods. (*Ar.* 246.)

(2.) Mr. Robert Patterson, in 1790, recommended placing earth connections in a kind of well filled with charcoal or embers. (*Do.* 247.)

(3.) It is a mistake to suppose that watertight cisterns make good earths. (*Do.*)

(4.) If the soil is humid, the metal soon rusts, but well-burned charcoal preserves the iron. (*Do.*)

(5.) Ordinary dry charcoal is found not to be a good conductor of "fulminating matter." (*Do.* 254.)

(6.) "A faulty termination of the earth connection is of all else the most common and frequent blunder in relation to lightning conductors that is made." (*Mann*, 1875, 536.)

(7.) A moist earth contact is recommended. (*Do.*)

(8.) "All competent electrical engineers are now keenly alive to the automatic electrolytic action that is apt to take place in the earth contacts of a lightning conductor." (*Do.* 537.)

(9.) "An earth contact of 1,000 square mètres (1,196 square yards) has been fixed by the best French authorities as sufficient for all practical purposes for a conductor of copper that is 1 centimètre ($\frac{1}{10}$ inch) square." (*Do.*)

(10.) M. Callaud proposes to use with this $2\frac{8}{10}$ bushels of broken coke. (*Do.*)

(11.) Occasionally a bore of 4 or 5 inches diameter, and 16 or 20 feet deep, into damp soil should be made for the insertion of the earth connection. (*Do.* 538.)

(12.) At the Hôtel de Ville, Brussels, the earths of the lightning rods are as follows :—At 3 feet from the ground the rods enter a box of galvanised iron filled with zinc poured in molten, out of which issue three bundles of iron rods to form earth connections. One bundle is carried beneath the pavement into a cast-iron cylinder,

II. D 13—22.

2 feet diameter and 8 feet long, sunk in a well hollowed in the ground, and providing a water contact. A second bundle is attached to the iron main of the gas service of the town. The third is attached to the iron water service of the town. The entire earth contact is about 330,000 square yards. (*Do.* 1878, 332.)

(13.) An earth contact of large extent is needed; the best is a gas or water main; otherwise a trench in moist ground, 20 feet long, packed with gas coke, should be used. (*Do.* 335.)

(14.) Where the earth is dry the contact must be larger. (*Do.*)

(15.) Conductors may be either connected with iron gas or water mains, buried in coke, attached to metal in moist earth, or carried down wells. (*Preece*, 347.)

(16.) At Jersey, the conductor of a church was found broken off 2 feet from the ground, and it had been so for years. (*Do.*)

(17.) At Lydney, the earth connection of the conductor of the church consisted of an iron gas-pipe leaded into a loose stone resting on a stone pavement. (*Do.*)

(18.) At Llandaff Cathedral, the conductor, a small copper stranded wire, fixed by galvanised iron wall hooks, was found corroded and eaten away, 1 foot below the surface, by electrolytic action, the surfaces at breakage clearly denoting the action of a current. (*Do.* 354.)

(19.) The potential of the atmosphere increases on ascending. Thus there is a current between the rod at the top of a spire and the earth. The result is electrolytic action at the junction with the ground. (*Do.*)

(20.) This is mitigated by making earth with as large a mass as possible. (*Do.*)

(21.) Conductors should invariably be continued through "light dry soil, such as shingle and sand," to soil which is permanently damp. (*W. O.* 1875, 4.)

(22.) Good earth connections are most important. Conductors should be led if possible into springs, wells of water, or ground permanently wet. (*Do.* 38.)

(23.) The sea is a good earth, also any body of water not enclosed in a watertight tank. (*Do.*)

(24.) Shingle, dry sand, and vegetable mould are bad earths. (*Do.*)

(25.) All large systems should have several earths, "so that should one be defective, the discharge may be effected through the other." (*Do.*)

(26.) Conductors should be led into moist ground by trenches 18 inches below the surface. (*Do.* 39.)

(27.) Not less than 30 feet of metal should be in contact with moist earth. (*Do.*)

(28.) A flow of water from the down-pipes of the roof should be led if possible over the earths. (*Do.* 40.)

(29.) Special precautions are necessary in rocky and dry soils. The trenches should extend 30 to 120 feet from the foot of the conductor. (*Do.* 41.)

(30.) Earth connections in trenches may be of railway or old iron. (*Do.* 42.)

(31.) The trenches should be filled with cinders, or with coal ashes. (*Do.*)

(32.) Water-pipes make good earths, but not gas-pipes, on account of their liability to be fused. (*Do.*)

(33.) In 1846, the War Office Regulations prescribed covered cisterns in preference to open trenches as being less liable to evaporation, and the cisterns were ordered to be kept full of water. (*Do.* 1846.)

(34.) Conductors should terminate below in damp or porous soil. (*Do.* 1858, 8.)

(35.) If the soil is dry, radiating trenches should be cut 30 feet long and 18 inches or 2 feet deep, and either the conductor itself or old iron chain carefully connected to its foot should be laid therein, and the trenches should be filled to a depth of 12 inches with coal ashes or other carbonaceous substance. (*Do.*)

(36.) Surface drainage should be led over the trenches. (*Do.* 9.)

(37.) Tanks are useless. (*Do.* 10.)

II. D 38—44.

(38.) Any watertight tanks in existence should be replaced by the above-named arrangements. (*Do.*)

(39.) As regards the use of iron earth connections with copper rods, "the contact of copper with iron will occasion a rapid oxidation of the latter metal." (*Franc. Mich. Tel.* 44.)

(40.) The rod is usually led into a well, and ends in two or three ramifications to connect better with the soil. (*Gan.* 833.)

(41.) If there is no well, a hole should be dug to a depth of 6 or 7 yards, and the foot of the rod introduced, the hole being filled up with wood ashes. (*Do.*)

(42.) In the case of the earths of telegraph wires "where there are neither gas nor water pipes, an earth plate is used; a plate of copper buried upright in a narrow trench filled with smith's ashes or wood charcoal, the object being to expose as large a surface as possible." (*Cull.*)

(43.) Professor Pouillet's Committee in their report, dated 19th February, 1855, stated:—There should be never-failing connection on the part of lightning rods with water or moist earth, and to insure this, the earth connections should be divided into two arms, the first "going very deep into the ground, into perennial water," and the second "running nearer the surface." (*And.* 81.)

(44.) In explanation of the above, Professor Pouillet wrote:—"After a long continuance of dry weather it often happens that the lightning-bearing clouds exert their influence only in a feeble manner on a dry soil, which is a bad conductor; the whole energy of their action is reserved for the mass of water which, by percolation, has formed below it. It is here that the dispersion of the electric force takes place." . . . "The case is entirely different when, instead of dry weather, there have been heavy rains moistening the earth thoroughly up to the surface. It is the latter now that is the best, because the nearest, conductor of the electric force, which will not go to the more permanent sheet of water lying more or less deep in the ground, if there is moisture above it." (*Do.*)

(45.) The French prefer a moist soil to mere water contact. (*Do.* 131.)

(46.) M. Callaud uses for an earth contact a galvanised iron grapnel placed between two layers of charcoal. (*Do.*)

(47.) "Probably in nine cases out of ten wherever a building provided with a conductor is struck by lightning, it is for want of good earth." (*Do.* 198.)

(48.) Franklin, in a report dated 21st August, 1772, regarding lightning rods for the Government powder magazines at Purfleet, proposed to dig a well at each end of each magazine, "in or through the chalk so deep as to have in it at least 4 feet of standing water." (*Do.* 199.)

(49.) "To dwell too largely upon the importance of leading all lightning conductors down into moist earth, or as technically called 'good earth,' would be scarcely possible." (*Do.* 198.)

(50.) The designing of earth connections should always be intrusted to experts. (*Do.* 210.)

(51.) "As regards the means of obtaining a good earth connection, the first and in all cases most preferable is to lay the conductor deep enough into the ground to reach permanent moisture." (*Do.*)

(52.) When the quantity of moisture is deficient or doubtful, "it will certainly be advisable to spread out the rope so as to run in various directions, similar to the root of a tree, likewise in search of moisture." (*Do.*)

(53.) "To protect any structure of great extent, it is absolutely necessary to bring the conductor or conductors deep enough into the earth to reach water." (*Do.* 212.)

(54.) In the case of a powder magazine standing on dry soil, and with no stream or permanent moisture near, "nothing remains under these circumstances but to multiply the lines of underground connection to the utmost extent."
. . . . "Still it must never be forgotten that absolutely good earth, in reference to lightning conductors, means moisture or water." (*Do.* 215.)

II. D 55, 56; E 1—5.

(55.) Iron water mains are recommended, if permanent moisture cannot be obtained. (*Do.* 216.)

(56.) Franklin suggested lead for earth connections as being less liable to consume with rust than iron. (*Frank.* 429.)

(E.) The Application of Rods.

(1.) "The case of a ship is very different from that of a building. A ship is a prominent object, generally a conductor, situated upon a plane, the sea. It thus, if a thundercloud passes near it, at once reduces the line of resistance between the sea (inner coating) and the cloud (the exterior coating of the condenser), determining discharge." . . . . "On the other hand, buildings form but an insignificant feature in the large irregular area exposed to induction from a charged cloud. Trees and buildings take but a portion of the charges which, in the case of ships, have fallen in their whole intensity upon them." (*Preece*, 342.)

(2.) A chimney lined with a thick layer of soot, up which a current of heated air and volumes of smoke are ascending, and terminated with a mass of metal (the grate), is an excellent but dangerous conductor, for it ends in the room and not in the earth. Hence so many indoor accidents, and hence the duty of every householder, particularly in exposed situations, to protect himself and his family. (*Do.* 345.)

(3.) Many houses are already protected by the lead-work of roofing and the iron rain-pipes descending to the drains. (*Do.* 348.)

(4.) At a house at Painswick, near Stroud, Gloucestershire, a 2-inch wrought iron drain-ventilation pipe is used as a conductor, carried 8 feet above the highest chimney, surmounted by a copper vane and point, and connected with the lead of the roof. (*Do.*)

(5.) For a sum of £2 an intelligent man can protect his house. (*Do.*)

II. E 6—15.

(6.) Professor Abel, F.R.S., having had considerable experience of conductors attached to powder magazines, was frequently astonished at the amount of complication introduced, and fully agreed with Mr. Preece as to the use of galvanised iron wire for conductors (see B. 27—29), and he importance of a thoroughly good permanent earth. (*S. T. E.* 357.)

(7.) Many accidents result from the want of a good earth connection to chimneys and fireplaces. (*Galt, S. T. E.* 358.)

(8.) Well-built houses are generally protected from lightning by the lead water-pipes and ridges about them, and it is from that cause that so few accidents to well-built houses are heard of. (*Lat. Clark, S. T. E.* 373.)

(9.) Underground magazines are usually in dry soil. The main underground magazines of works of defence should be fitted with conductors. This is unnecessary with small expense underground magazines. Casemated batteries with magazines in their basements should have conductors. (*W. O.* 1875, 5—8.)

(10.) Flagstaffs of coast batteries should have conductors. Iron shields of batteries must be connected with conductors. (*Do.* 9.)

(11.) Asphalte and concrete roof coverings are non-conductors. (*Do.* 10.)

(12.) A building of uniform height should have a pointed rod, 5 feet above it, at intervals of 45 feet along its length. If of iron, the point should be gilt. (*Do.* 29.)

(13.) Buildings not more than 20 feet long to have one vertical conductor at the end, with a point 5 feet above the roof, and a horizontal conductor along the ridge. (*Do.* 30.)

(14.) If 20 to 40 feet long, there should be one vertical conductor in the centre, with a horizontal conductor along the ridge. (*Do.* 31.)

(15.) If exceeding 40 feet long, there should be two vertical conductors, and if exceeding 100 feet, three. (*Do.* 32.)

II. E 16—24.

(16.) A diagram is given as an illustration of the manner of defending a large powder magazine. The building has four parallel ridged roofs with gable ends, and the following system of rods:—

> (1.) 8 copper-pointed terminal rods, one at each gable summit.
> (2.) 4 horizontal copper ridge pieces, joining the terminal rods.
> (3.) 16 sloping copper gable end pieces, leading from the terminal rods to the gutters.
> (4.) 2 copper eaves-gutters at the sides.
> (5.) 3 lead valley-gutters between the roofs.
> (6.) 8 copper down-pipes from the gutters to earth. (*Do.* 33.)

(17.) All parts of a building of marked elevation should be fitted with conductors. (*Do.* 34.)

(18.) Where several conductors are used in a building, they should be connected horizontally. (*Do.* 35.)

(19.) "All metal surfaces, whether of lead, copper, or iron, in ridges, roofs, gutters, or coverings to doors or windows, should be connected with the conducting system." (*Do.*)

(20.) Materials of which buildings are composed are for the most part conductors. (*Harr.* 94.)

(21.) It can be seen by experiment how walls of buildings are conductors of electricity. (*Do.* 9.)

(22.) A wall is a protection to a house, and the rod should be outside a building, and not inside. (*Ar.* 249.)

(23.) It having been observed that the minute dust of gunpowder, liable to lodge on the projections and ledges of powder magazines, was a source of danger, Toaldo, in 1776, suggested placing conductors at upright masts some feet distant from the magazines. (*Do.* 251.) [*See note to* G. 45.]

(24.) As many conductors as terminal rods are needed. (*Do.* 247.)

(25.) It is advantageous to connect the bases of the rods along the ridge of the roof, also any metals in the roof and parapet. (*Do.*)

(26.) The use of insulating substances between the rod and the house is now (1855) nearly given up, being recognised as unnecessary and costly. (*Do.*)

(27.) The object of rods "is to permit a free neutralisation of the electric forces, and thus, as it were, to afford a ready outlet to a violent agency that may do mischief to an indefinite amount if not provided with such means of escape." (*Nels. A. M.*)

(28.) In Gay Lussac's report to the French Academy in 1823, it was held that "all large metallic masses contained in any building should be brought into metallic communication with the main system of conductors, and that there was no need whatever for the employment of insulating supports in attaching the lightning rod to the structures that it is intended to defend." (*Mann*, 1875, 538.)

(29.) M. Callaud, in his recent treatise on "Paratonnerres," does not agree with this. He adopts insulating supports, and he contends that the rod itself ought to be quite sufficient, and that the metallic connections are superfluous when it is efficient, and dangerous when it is not. (*Do.*)

(30.) Lightning protection should not be carried near to gas-pipes smaller than 1 inch in diameter, nor to any made of soft metal. (*Do.* 1878, 332.)

(31.) "The lead roofing and all masses of metal in the line of the probable discharge should be connected with the rods." (*Do.* 347.)

(32.) Conductors need not always be outside a building. (*Do.*)

(33.) In America, gutters, rain-pipes, and other metal surfaces are much utilised as conductors. (*And.* 134.)

(34.) The large iron mineral oil tanks are protected by conductors. (*Do.* 138.)

(35.) "The first point in designing the protection of a

II. E 36—41.

building will be clearly to ascertain what path the lightning will take on its course from the clouds to the earth." (*And.* 141.)

(36.) Mr. Anderson gives (but without an estimate) a design "for the protection of a large detached mansion by means of a multiplication of short points, or terminal rods, fixed on all the prominent features of the building. The conductor is carried along the ridges in every direction, and down the edges of the roof at each gable. Generally it is sufficient to have two descending conductors, but occasionally the conformation of the building, or the nature of the ground, renders necessary the use of even more." (*Do.* 150.)

(37.) Franklin was at first in favour of having lightning rods inside buildings. This plan was adopted in France and on the Continent, but was soon abandoned. (*Do.* 158.)

(38.) The practice of insulating the rod from the building "is not only useless, but positively dangerous." (*Do.*)

(39.) There is a necessity "of leaving the design and erection of lightning conductors to those persons who have made a thorough study of the subject, since the work is by no means so free from complexity as is commonly supposed." (*Do.* 177.)

(40.) "The idea that persons may construct their own conductors is left aside altogether as absurd." (*Do.* 216.)

(41.) As tending to show how little the application of rods to buildings has been associated in England with the architect's profession, it may be mentioned that the following well-known architectural works of reference apparently contain no allusion to the subject of defence from lightning, viz. :—

(α) Nicholson's Architectural Dictionary, 1819.
(β) Bartholomew's Specifications for Practical Architecture, 1840.
(γ) Gwilt's Encyclopædia of Architecture, 1842.

II. E 42; F 1—6.

(δ) Weale's "Dictionary of Terms used in Architecture," "Building," "Engineering," &c., 1858—9.

(ε) Kerr's "Gentleman's House," 1865.

(42.) An iron building would be wholly free from damage by lightning. We must, therefore, endeavour to bring the general structure "as nearly as may be" into the same state as if it were all of metal. (*W. O.* 1858, *App. A.* 3, 4, 5.)

(F.) INSPECTION OF RODS.

(1.) The state of the earth should be tested by means of galvanometers. (*Mann*, 1875, 537.)

(2.) M. de Fonvielle has recommended an arrangement of a short circuit wire, with a galvanometer, to be fixed to each separate conductor, so that an examination can always be made. (*Do.*)

(3.) A convenient form of galvanometer for testing conductors has been devised by Mr. R. Anderson. (*Do.* 1878, 339.)

(4.) Lightning rods should be periodically examined as to points, continuity, and earth, and tested with a galvanometer and current. (*Preece*, 347.)

(5.) Frequent inspections are needed. (*W. O.* 1875, 43.)

(6.) "Most of the lightning conductors in Paris have been neglected for so many years that they are positively dangerous, instead of being useful protecting apparatus. When examined with a strong magnifying glass, the points of the stems are blunted or burnt (a criterion of the bad conduction of the communications), and the points have fallen from several, or rather, having been badly joined, the solder has failed, and they only hold together by the pins; the vibrations of the stem when agitated by the wind have worn away the connections, so that a great number are easily shaken by the hand. The contact is very bad,

II. F 7—15.

consequently the preventive effect of the apparatus is absolutely null. But these great deteriorations are not confined to the stems, since the point of juncture of the conductor to the base of the stem is almost everywhere in a deplorable state. This juncture I have always found to have been made with a strap or iron collar, whose pieces are rusted, so as to render the electric communication next to nothing." (*Franc. Mich. Tel.* 44.)

(7.) The joints of the above rods being badly arranged, have been affected by the dilatation and contraction due to variations of temperature. (*Do.*)

(8.) "Where the conductor penetrates the soil it is not covered with any protecting substance; so that the alternation of dryness and moisture in the soil deeply corrodes the iron and ultimately cuts it through." (*Do.* 45.)

(9.) Without periodical tests, "there is really no trustworthy security of protection in lightning conductors." (*And.* 61.)

(10.) The French committee who reported on the 14th January, 1867, stated that conductors, to be efficient, should be regularly inspected, at least once every year. (*Do.* 84.)

(11.) The conductors of Westminster Palace have never been tested since they were fixed. (*Do.* 120.)

(12.) The inspection of conductors after once they are put up is generally overlooked in this country. (*Do.* 218.)

(13.) The regular inspection of lightning conductors, as yet unknown, or all but unknown, in England, has been for a long time in practice in several States of Continental Europe, among them Germany and France. (*Do.* 222.)

(14.) "There is really nothing else to make a lightning conductor a safe protection under all circumstances, and at all times, but regular, constant, and skilful examination." (*Do.* 226.)

(15.) The writer recently tested 16 "earths" of the lightning conductors of some large powder magazines with one cell of a Le Clanché battery and an ordinary vertical detector galvanometer, all the arrangements being made by a practised telegraphist. The "quantity" coil of the galvanometer gave no deflection at all for any of the "earths," and the "intensity" coil gave about 2° for those tried with it. On a subsequent occasion one of the same "earths" was tested with 3 Le Clanché cells, and a 3-coil galvanometer, by the 10 coil of which a deflection of from 25° to 35° was registered.

(G.) The Protective Powers of Rods.

(1.) The Section of Physic of the French Academy, consulted in 1823 by the Minister of War, considered that a rod protected a circular space whose radius was equal to twice its height. (*Ar.* 237.)

(2.) Arago knew of no case in which pointed conductors had failed to preserve buildings from damage, in which there had not been palpable errors of construction. (*Do.* 264.)

(3.) In 1765, the physicist, Nollet, opined that rods attracted lightning, and occasioned it to strike a house more frequently than would otherwise be the case. (*Do.* 265.)

(4.) Another physicist, Wilson, supported this view. He and Nollet thought that rods were, on the whole, more dangerous than useful. (*Do.*)

(5.) A definite radius of protection could not always be assigned to rods; and experience showed that their influence was only in furnishing "an easy line of conduction" to the discharge. (*Harr.* 117.)

(6.) The cases in which buildings with rods had been damaged, bore only "a small proportion to the great mass of instances in which lightning falling on buildings has struck on the conductors attached to them." (*Do.* 120.)

(7.) In 1838, the East India Company, owing to the

II. G 8—14.

representations of their scientific officers, ordered the rods to be removed from their powder magazines and other public buildings. (*Do.* 176.)

(8.) In a work published in 1829 by a Civil Engineer in the Government service, entitled "Three Years in Canada," the following passage occurred:—"Science has every cause to dread the thunder rods of Franklin; they attract destruction, and houses are safer without them than with them." (*Do.* 177.)

(9.) Harris says that the theory just mentioned is not warranted by any sound argument drawn from experience. (*Do.* 184.)

(10.) It is inferred from "the fact of so many buildings being repeatedly struck by lightning before they are furnished with lightning rods, and so seldom struck afterwards, and from the fact of lightning having seldom, if ever, been observed to fall in an explosive form upon buildings involving pointed metallic conductors in the construction," that the rods had "rapidly neutralised the electrical state of the air and so prevented the occurrence of a dense explosion." (*Do.* 189.)

(11.) In the experience of nearly half a century, "not a single case can be adduced in which a lightning rod, in the act of transmitting a heavy charge of lightning, had thrown off a lateral explosion on semi-insulated masses near it." (*Do.* 201.)

(12.) Faraday had stated (previously to 1843) that he was "not aware of any phenomenon called lateral discharge which is not a diversion of the primary current." (*Do.* 208.)

(13.) "The few accidents on record can scarcely be urged as an objection to the general principle, especially when we take into consideration the great number of lightning rods set up in various parts of the world." (*Do.* 224.)

(14.) "The lightning stroke is certainly more likely to fall where a lightning conductor, of whatever kind, is

II. G 15—23.

placed, than it would be if there were no such appliance." (*Mann*, 1875, 333.)

(15.) "The old dogma, that a conductor does not attract electricity, is open to modification. The induction draws a strong charge to the top of the rod, and thus brings about a stronger tendency to discharge." (*Do. Times*, 23 *Nov.* 1877.)

(16.) It is doubtful if a conductor of insufficient size is better than none at all. (*Do.* 1875, 533.)

(17.) Attention is called by Dr. Mann to the cases of tall chimney-stacks with conductors being struck; and this he attributes to the frequently insufficient size of the latter, which causes the discharge to leap through the brickwork to the soot-covered surface within. (*Do.*)

(18.) When the point is blunt, or the earth contact bad, rods attract lightning. There is no attraction in a well-constructed conductor. (*Do.* 1878, 330.)

(19.) It is contended that the usual view of a conical space being protected by a conductor is not trustworthy; and the only absolute protection is to cover the entire structure with intermeshed lines of defence. (*Do.*)

(20.) After the city of Pietermaritzburg, in Natal, had been largely supplied with pointed lightning conductors under Dr. Mann's fostering influence, "the actual discharge of violent lightning strokes within the area of the town became almost unknown." (*Do.* 1875, 335.)

(21.) The radius of cone protected by a conductor should be taken at half the height of the conductor. (*Preece*, 348.)

(22.) "The methods that have been adopted for protection, based upon the damage inflicted on ships, have probably led to the adoption of unnecessarily costly and superfluous measures to protect buildings and instruments." (*Do.* 342.)

(23.) "As regards ships, the method adopted by the great authority, Sir William Snow Harris, has proved itself so efficient and perfect that no improvements have been required, nor can any well be suggested. . . . Such

II. G 24—32.

vessels as have been struck have been invariably unprotected." (*Do.* 344.)

(24.) It is "sufficiently correct for practical purposes" to take the protective radius of the conductor the same as its height above the ground; but this cannot always be relied on. (*W. O.* 1875, 26.)

(25.) "Conductors of themselves have no attraction for lightning, which seeks them only on account of the facility they afford for the combination of the opposed states of the electricity of the clouds and the earth beneath them, separated by the atmosphere, which is a bad conductor." (*Do.* 27.)

(26.) "A lightning conductor, buildings, trees, or any object on the surface of the earth is only to be regarded as diminishing the resistance due to the air." (*Do.*)

(27.) "When an electrified cloud is passing over the earth, and its potential is just counteracted by the resistance of the air, a body, however small, which reduces the resistance will cause a discharge." (*Do.*)

(28.) "Even a change in the nature of the soil over which the cloud is passing may have this effect on it." (*Do.*)

(29.) "It is very frequently produced by a fall of rain." (*Do.*)

(30.) An angle of a building may receive a discharge, while another angle is provided with a conductor. Important buildings containing explosive materials should have every prominent elevated part provided with a conductor. (*Do.* 28.)

(31.) "A conducting rod, in whatever way it may be applied, is to be considered merely as a means of perfecting the conducting power of the whole mass so as to admit of intense discharges of lightning being securely transmitted, which otherwise would not pass without intermediate explosion and damage." (*Do.* 94.)

(32.) The idea of lateral discharge from conductors is absurd. (*W. O.* 1858, *App. A.* 8.)

(33.) Masonry, ships' masts, and lightning conductors transmit a certain quantity of electricity without explosive action. The conductor relieves the wood and the masonry. (*Do. App. B.* 2.)

(34.) Experience having shown, soon after the introduction of conductors, that by their use lightning became without effect, "it was thought that storms might be dissipated if a sufficient number of *paratonnerres* were raised, so as to neutralise the electricity of the atmosphere." The houses at Zurich are studded with *paratonnerres*. (*Kaem.* 353.)

(35.) M. Viollet-le-Duc, the distinguished French architect, considers it prudent to put up two rods on a typical country house (the design and construction of which is the subject of his book), since the recognised theory is that "lightning conductors only protect the points enclosed in a cone of which they are the summit." (*Violl.* 253.)

(36.) "Between ourselves, physicists are not quite agreed respecting the effects of the electric fluid, the relative efficiency of conductors, and the precautions to be used in putting them up." (*Do.*)

(37.) "I rely on my own experience which has proved to me that no building, however exposed, has been struck by lightning when the lightning rods were numerous, made of good conductors, put in communication with each other, and with their lower extremity dipping in water or very damp earth." (*Do.*)

(38.) M. Viollet-le-Duc does not see the advantage of using insulators. (*Do.*)

(39.) "Projections such as trees, spires, lightning conductors, lead off the negative electricity of the ground quickly, thereby diminishing its tension, and conduct the electricity of the clouds to the earth without violence." (*R. E. A.* 56.)

(40.) As regards the lightning conductor, "the advantage gained by it consists, not in protecting the building in case of a discharge by allowing a free passage for the

II. G 41—46.

electric fluid to escape to the earth, for it is but a poor protection in such a case; but in quietly and gradually keeping up the communication it tends to maintain the electric equilibrium, and thus to prevent the occurrence of a discharge." (*Buch.* 300.)

(41.) The result of the report of the committee of the French Academy in April, 1823, that rods would protect a circular area of a radius double their height, led to rods of enormous height being erected. (*And.* 77.)

(42.) The French Committee of December, 1854, reported through M. Pouillet that the theory of a fixed area of protection was inadmissible. (*Do.* 78.)

(43.) They gave as their reason the varying shapes of buildings and materials of construction, and said "it is clear, for example, that the radius within which the conductor gives protection, cannot be so great for an edifice, the roof, or upper part, of which contains large quantities of metal, as for one which has nothing but bricks, woods, or tiles." (*Do.* 79.)

(44.) They also said "a lightning conductor is destined to act in two ways. In the first place, it offers a peaceful communication between the earth and the clouds, and by virtue of the power of points the terrestrial electricity is led gently up into the sky to combine with its opposite. In the second, it acts as a path by which a disruptive discharge may find its way to the earth freely." (*Do.*)

(45.) The French Commission who reported on the 14th January, 1867, as regards powder magazines, stated "that the best protection against lightning would be afforded by the most substantial metal rods, made of iron, surrounding a building on all sides, and passing deep into the ground."* (*And.* 84.)

(46.) In a work entitled "Three Years in Canada," published in London in 1829, Mr. F. McTaggart, C.E.,

* Guillemin, in his "Application of Physical Forces," says that these rods are proposed to be attached to wooden masts separated from the magazines.

II. G 47—50.

wrote about lightning rods, "Were they able to carry off the fluid they have the means of attracting, then there could be no danger; but this they are by no means able to do." (*Do.* 92.)

(47.) The general presumption in France is that "a terminal rod will protect effectually a cone of revolution of which the apex is the point of the rod, and the radius of the base a distance equal to the height of the said rod above the ridge multiplied by 1·75." (*Do.* 126.)

(48.) "The function of a lightning conductor is twofold. In the first instance, it operates as a medium by which explosions of lightning, or, to speak more accurately, disruptive discharges of electricity, are led to the earth freely. In the second instance, the conductor acts as a means whereby the accumulation of electricity existing in the atmosphere is quietly drawn off and carried noiselessly into the earth, and dissipated in the subterraneous sheet of water beneath it." (*Do.* 142.)

(49.) Professor J. Clerk Maxwell, F.R.S., has proposed a system of lightning protection without points or earth connections. He says, "What we really wish to prevent is the possibility of an electric discharge taking place within a certain region—say in the inside of a gunpowder manufactory. . . . An electrical discharge cannot take place between two bodies unless the difference of their potentials is sufficiently great compared with the distance between them. If, therefore, we can keep the potentials of all bodies within a certain region, equal, or nearly equal, no discharge will take place between them." (*Do.* 164.)

(50.) To do this he proposes "to enclose the building with a network of good conducting substance. For instance, if a copper wire, say No. 4, B. W. G. (0·238 inches in diameter), were carried round the foundation of the house, up each of the corners and gables, and along the ridges, this would probably be a sufficient protection for an ordinary building against any thunderstorm in this climate." (*Do.*)

II. G 51—54.

(51.) In respect of lightning rods generally, Mr. Anderson says, "Subject to the constant effects of moisture, to wind, ice, and hailstorm, there is always a possibility of the slender metal strips being damaged so as to interrupt their continuity and thus destroy the free passage of the electric force. Instances have happened in which the damage done was so slight as to be scarcely visible, and still to destroy the efficacy of the conductor." (*Do.* 219.)

(52.) The improved drainage now going on everywhere constitutes a serious danger to lightning conductors, since the moisture is being sucked out of the ground. (*Do.*)

(53.) "Constant alterations in the interior of buildings, private residences, as well as public edifices, may serve to destroy the efficacy of a conductor which was originally good even to perfection. Thus a roof may be repaired, and lead or iron introduced where it was not before; or clamps of iron may be inserted in the walls of houses, to give them greater strength; or in fact any changes may be made which bring masses of metal more or less in proximity to the conductor.. . . . There are hundreds of instances to prove that changes made in buildings such as the addition of a leaden roof without, or the iron balustrade of a staircase within, diverted the current of the electric force from the conductor on its way to the earth, originally well provided for." (*Do.* 220.)

(54.) "A lightning rod protects a conic space whose height is the length of the rod, whose base is a circle having its radius equal to the height of the rod, and whose side is the quadrant of a circle whose radius is equal to the height of the rod." . . . "There are many cases where the pinnacles of the same turret of a church have been struck where one has had a rod attached to it; but it is clear that the other pinnacles were outside the cone, and therefore, for protection, each pinnacle should have had its own rod. It is evident also that every point of a building should have its rod, and that the higher the rod, the greater is the space protected." (*Preece, Tel.* 15/12/80.)

II. G 55, 56.

(55.) A conductor of insufficient sectional area, "if connected with the earth, would protect a house from injury, though it might itself be destroyed." (*Lat. Clark, S. T. E.* 374.)

(56.) Mr. Graves recommends erecting numerous lightning conductors everywhere, especially upon high hills and high buildings, in order, so far as possible, to prevent the occurrence of thunderbolts. (*Grav. S. T. E.* 413.)

NOTE.—Whilst this work was passing through the press, the Report of the "Lightning Rod Conference" (London: Spon, 1882) was published. The Report is dated 14th December, 1881, and is signed by delegates from the Meteorological Society, the Royal Institute of British Architects, the Society of Telegraph Engineers and of Electricians, and the Physical Society. The Report says, "A lightning conductor fulfils two functions: it facilitates the discharge of the electricity to the earth, so as to carry it off harmlessly, and it tends to prevent disruptive discharge by silently neutralising the conditions which determine such discharge in the neighbourhood of the conductor." The points must be "high enough to be the most salient features of the building, no matter from what direction the stormcloud may come." The following instructions are given:—Tops of rods to be blunt, but 1 foot below them a copper ring with three or four sharp copper points to be fixed—points to be either platinised, gilded, or nickel plated; the best material for rods is copper; the best form is a copper rope $\frac{1}{2}$" diameter, or a copper tape $\frac{3}{4}$" × $\frac{1}{8}$"; joints to be always soldered; iron rods to be painted (except at points) even if galvanised; insulators not to be used; gas and water mains to be utilised as earths; earth plates to be of the same metal as the rods, to be at least 9 square feet in area, and to be sunk in holes so deep that the earth around them is always moist; in rocky sites, besides earth plates, 3 or 4 cwt. of iron to be buried at the foot of rod; the space protected by a rod is a cone with a base having a radius equal to the height of rod; rods must periodically be examined visually and tested electrically; internal masses of metal (except soft metals and gas-pipes) to be connected to earth, or to the rod; external masses of metal to be connected to each other and to earth direct, or to the rod.

III. CHAPTER III.—SOME INCIDENTS OF THE ACTION OF LIGHTNING. 86

No.	Time.	Place.	Object.	Particulars.	Lightning rods present.	Metals present.	Results.	Authority
1	12 Sept. 1747.	Near Toulouse, France.	A woman named Bordenave.	Clear sky, except a small round cloud.	..	..	Killed.	Ar. 9.
2	30 July, 1764. (5.30 a.m.)	Near Pithiviers, France.	Elm tree.	Bright sunshine with one small cloud.	..	..	Strip of bark 3 in. wide torn off.	Ar. 9.
3	26 Aug. 1827. (Vespers.)	Admont convent chapel, Austria.	Two priests in the choir.	Chapel with belfry and cross in a valley.	..	..	Killed. (Height of cloud only 92 feet.)	Ar. 19.
4	1 May, 1700.	Church on Mt. St. Ursula, Styria.	Seven persons sitting in church.	On the summit of a high mountain.	..	..	Killed. (Thunderstorm and clouds half way up the mountain; at summit sky serene and sun brilliant.)	Ar. 40.
5	19 Apl., 1827.	Gulf Stream, North Atlantic.	Packet Ship *New York*.		Iron-pointed rod ½ in. diameter, 4 ft long; thence iron chain, 130 ft. long, to sea; links ¼ in. diameter.	..	Rod struck. Few inches near point melted. All chain, except 3 feet, dispersed. (Sea boiled like as if acted on by a volcano. Three columns of water twice shot up into the air and fell back foaming.)	Ar. 71, 96. Harr. 109.

No.	Date	Place	Object struck	Circumstances	Conductor	Other details	Effect	Ref.
6	1787.	Philadelphia, North America.	House.	..	Benjamin Franklin's.	Copper rod (at summit) $1\tfrac{3}{10}$ in. thick at base, $9\tfrac{1}{2}$ in. long.	Rod struck and melted.	Ar. 72.
7	Aug., 1777.	Cremona, Italy.	High tower.	..	..	Weathercock.	Weathercock struck, but rod of it, $\tfrac{1}{2}$ inch thick, showed no trace of fusion.	Ar. 74.
8	12 July, 1770.	Philadelphia, North America.	House.	..	..	$\tfrac{1}{2}$ in. round iron rod.	Rod struck, but not injured.	Ar. 74.
9	18 June, 1782.	Stoke Newington.	House.	..	..	Wire of bedroom door drop bolt.	Wire shortened in length of $16\tfrac{1}{2}$ feet by 2 or 3 inches.	Ar. 75.
10	3 Sept., 1789.	Lord Aylesford's park, England.	A man.	Sheltering under an oak tree.	..	..	Killed. Perforation in ground at man's walking stick (on which he was leaning) 5 in. deep and $6\tfrac{1}{2}$ in. diameter. Soil found blackened 10 inches deeper, and 2 inches deeper still quartzose stone showed traces of fusion.*	Ar. 78.
11	14 Apl., 1718.	Gouesnon, near Brest, France.	Church.	..	..	..	Walls overturned. Some stones thrown to 55 yards distance.	Ar. 86.
12	Jan., 1762.	Breag, in Cornwall.	Church.	..	..	..	A stone of 336 lbs. thrown 60 yards. Another stone found 400 yards off.	Ar. 86.

* Lord Aylesford erected a small pyramid at the tree with a warning inscription.

III. SOME INCIDENTS OF THE ACTION OF LIGHTNING.—(*Continued.*)

No.	Time.	Place.	Object.	Particulars.	Lightning rods present.	Metals present.	Results.	Authority.
13	29 June, 1763. 20 June, 1764.	Autrasme, near Laval, France.	Church.	With steeple, and a credence table of soft stone.	.. ..	Gilt picture-frames and decorations, and pewter flasks inside church.	On 29th June, 1763, gildings of frames and decorations blackened and flasks fused, and two holes drilled in credence table. All was repaired, and exactly the same damage was done again on 20th June, 1764.	Ar. 93.
14	10 Sept. 1841.	Peronne, France.	House.	..	..	..	The particular room again struck where 25 years before the poet Béranger had narrowly escaped death by lightning.	Ar. 93.
15	19 July, 1785.	Near Coldstream, just across the Tweed.*	A man.	Driving two horses in a cart.	..	Wheel tires of cart.	Man and horses killed. Cart overturned, holes found at spots where wheels had rested, and tires fused thereat. Hair of horses burnt, particularly on legs and under belly.	Ar. 98. Harr. 63.

* Thunderstorm in progress. Event seen by two witnesses who heard thunder but saw no lightning.

Church.	Timber spire 70 feet high.	..	Bell wire 20 feet long from clock to clapper.	Some stones at foundation torn up, tower injured, bell-wire melted, spire shattered and thrown to a distance.	Ar. 140.
House in Rue Plumet.	..	..	Gilt picture-frame, lantern, iron stove, iron articles in a box, all inside house.	House struck, and lightning picked out these metals.	Ar. 140.
Lord Tylney's house.	Grand reception going on. 500 persons present in 9 rooms.	..	Gilding in rooms, i.e. mouldings, cornices, rods, sofas, chairs, doorposts, bell-wires.	House struck, and all gilding mentioned found affected on the following day. Some bell-wires partly fused. No person injured.	Ar. 140.
Mr. Raven's house.	..	Pointed iron rod projecting above roof, and a small brass wire leading thence to an iron bar in ground.	A fowling-piece resting inside the house against the wall outside which the brass wire passed.	Rod struck. Wire melted down to level of top of gun-barrel, where wall was pierced, and the discharge passed by gun-barrel.	Ar. 140. Harr. 206.
Two soldiers loaning against the wall of a chapel.	Portion of a detachment who were taking shelter outside.	..	Massive iron bars at the wall inside of chapel.	The two men killed. Wall torn open close to position of bars. Rest of detachment uninjured.	Ar. 142.

III. Some Incidents of the Action of Lightning.—(*Continued.*)

No.	Time.	Place.	Object.	Particulars.	Lighting rods present.	Metals present.	Results.	Authority.
21	28 Aug. 1760.	North America.	Mr. Maine's house.	..	Iron rod. ..	..	Rod struck and partly fused. Holes made in ground below rod. Slight injury to foundations.	Ar. 145.
22	5 Sept., 1779.	Mannheim, Germany.	Saxon ambassador's house.	..	Iron rod. ..	..	Rod struck. A whirl of sand at its base.	Ar. 145.
23	17 July, 1880.	Liverpool.	Mr. Bellion, a watchmaker.	Standing at the door of his house in Park Road.	..	..	Killed. ..	Stand. 19/7/80.
24	24 Feb., 1774.	Rouvoir, near Arras, France.	Church.	..	..	..	Great blue stones forming pavement of porch beneath steeple lifted vertically, and steeple struck.	Ar. 176.
25	Summer, 1787.	Tacon, near Beaujolais, France.	Two persons sheltering under a tree.	..	..	..	Killed. Portions of their hair found in top of tree, and portions of *sabots* found among branches.	Ar. 175.
26	29 Aug. 1808.	Paris.	A workman in a pavilion with thatched roof.	Near L'Hôpital de la Salpêtrière.	..	..	Killed. Portions of man's hat found inserted in ceiling.	Ar. 175.

27	5 Nov., 1755.	Maromme, near Rouen, France.	Powder magazine.	Containing 800 barrels.	...	...	...	A rafter of roof split, and two barrels filled with powder broken to pieces, but powder not ignited. Ar. 185.
28	11 June 1775.	Venice.	St. Secundus Church.	With a tower, and basement used as a powder magazine.	...	...	...	Shelves torn off magazine. Powder cases overthrown. Tower struck. Powder not ignited. Ar. 185.
29	May, 1843.	Commerage, near Lamballe, France.	Oak tree.	...	...	...	...	Bark showed a slit narrowing upwards from base to upper branches. Woody fibres clearly torn from below upwards. Ar. 177.
30	1670.	Zirknitz.	Lake.	...	...	...	...	Struck. Afterwards 28 small cartloads of dead fish were removed from the surface. Ar. 189.
31	14 Sept. 1772.	Besançon, France.	Lake.	...	...	...	...	After a lightning stroke, surface found covered with fish. Ar. 189.
32	3 July, 1828 (night).	Birdham, near Colchester.	Wooden bedstead in a house.	With a person in bed.	...	...	...	Bedstead broken to pieces. Person uninjured. Ar. 190.
33	9 July, 1828 (night).	Great Houghton, near Doncaster.	Bed in a house.	With a person in bed.	...	...	...	Coverlet torn off bed. Person uninjured. Ar. 190.

III. SOME INCIDENTS OF THE ACTION OF LIGHTNING.—(*Continued.*)

No.	Time.	Place.	Object.	Particulars.	Lightning rods present.	Metals present.	Results.	Authority.
34	29 Sept. 1772 (night).	England.	Mr. Thomas Hartley.	In bed in a house.	..	..	Killed. Wife by his side uninjured.	Ar. 190.
35	27 Sept. 1819 (5 a.m.)	Conflens, Charente, France.	A servant-maid.	In bed in a house.	..	..	Killed. Body had a furrow between neck and right leg.	Ar. 190.
36	17 July, 1880.	Liverpool.	St. Philemon's church.	With belfry.	..	Bell (presumed).	Belfry destroyed.	Stand. 19/7/80. Ar. 194.
37	21 July, 1819.	Biberach Prison, Suabia.	A prisoner.	Chained by the waist.	..	..	Struck. The only man out of 20 present in hall who was thus chained; none of the others struck.	Ar. 194.
38	6 Aug. 1753.	A house in St. Petersburg.	Professor Richmann.	Whilst experimenting with a lightning rod.	..	..	Killed. His body accidentally provided a passage for the discharge over a gap purposely made in the rod for drawing sparks.	Ar. 232.
39	12 May, 1777.	Purfleet.	Government store-house.	Rod erected under direc-	Rod projecting 11 ft. above roof.	Iron clamps of coping-stones.	Struck at a point 24 ft. distant from rod,	Ar. 237. Harr.

Eastbourne.	Two men.	On the ground floor of Mr. Adair's house.	..	..	Killed. Much damage done on first floor; upper story and roof untouched.	Ar. 242.
Lausanne.	Cathedral.	With steeple.	..	Horizontal iron bar in steeple at two-thirds of height.	Iron bar struck. Portions of steeple above bar apparently not struck.	Ar. 242.
Toothill, Essex.	Windmill.	In repose.	..	Iron knob in middle of sail.	Knob struck. Upper portions of sail untouched.	Ar. 243.
Milan.	Cathedral.	With pinnacles.	Rod on pinnacle leading to a supposed well.	..	Pinnacle struck. Much damage to building near rod at various elevations. (Supposed well proved to be a tiled cistern.)	Ar. 247.
Genoa.	Lighthouse.	..	Rod supposed to lead into sea.	..	Rod struck, and broken in several places. (Found to lead into a cistern in rock.)	Ar. 248.
Charlestown, North America.	300 prisoners and warders.	In the prison. Most had hammers, files, guns, or pikes.	Three rods in good condition 18 ft. apart.	Large amount of iron in prison besides what men carried.	Sustained a violent shock, their muscular strength being weakened for several seconds; no permanent injury. Building apparently untouched.	Ar. 250

III. SOME INCIDENTS OF THE ACTION OF LIGHTNING.—(*Continued.*)

No.	Time.	Place.	Object.	Particulars.	Lightning rods present.	Metals present.	Results.	Authority.
46	16 Dec., 1852.	St. Anne D'Auray, France.	Seminary.	..	Rod on tower.	..	Rod struck and fused. "Broken at the part where, after having followed the contours of the cornice, it bent again to descend vertically to the ground."	Ar. 251.
47	23 Feb., 1829.	Bayonne, France.	Powder Magazine.	50 ft. by 36 ft., covered by thick vaulted masonry and a sloping roof with gable ends.	Rod 1$\frac{3}{10}$ in. diameter, projecting 22 ft., and carried outwards from the foot of the wall, at 2 ft. above the ground, on wood posts to which attached by lead plates, into a trench filled with charcoal, 33 ft. from wall.	All the metallic parts of roof in connection with rod. Lead plates on gable ends. Lead gutters. Metal rain-pipes. Iron cramps to coping-stones.	Rod struck. Point fused for $\frac{3}{4}$ inch in depth. Lead plates at wood posts torn. At corner of building distant from rod a lead plate at gable near an iron cramp torn out.	Ar. 252. Harr. 213.
48	Jan., 1814.	Plymouth Harbour.	Ship *Milford*.	..	..	..	Injured. The only vessel in harbour at the time without a rod.	Ar. 264.

49	Summ'r 1831.	Vallera, near Parma, Italy.	House (overlookd at less than 70 yards off by trees and village church tower).	Of Signor Melloni, a physicist.	Rod, "a pretty thick one", with a copper-gilt point, leading into a well always holding water.	...	Rod struck and much shaken. Point entirely fused. House received no injury. (Rod erected in 1830. Inhabitants did not remember house, trees, or church over having been struck prior to this.)	Ar. 264.
50	1794.	Martinique, West Indies.	H.M. Ship *Dictator*.	...	...	...	Struck by a "fire-ball."	Harr. 21.
51	28 July, 1842.	London.	St. Martin's Church.	During fine weather. Top of spire 200 ft. from ground. Stands 44 ft. high on an open cupola surrounded by open columns and arches. Dial-room below, then belfry, then clock-room; lastly ringing-chamber level with church roof.	...	(1.) Iron vane spindle, 27 ft. long, 4¼ in. square, projecting 12 ft. into the air. (2.) Gilt copper vane, 8 ft. by 6 ft. (3.) Gilt copper ball 1 in. thick, 2 ft. 9 in. diameter. (4.) Strong iron supporting cross-frame, weighing 12¾ cwt. (5.) Two others lower down form-	Masonry of spire so damaged as to leave it tottering. Joints all loosened. Two stones thrown out, and two dislocated. Wood framework of cupola shivered. Gilt letters and minute hand of dial blackened. Brass screws of rod (12) slightly fused. Small copper wire of clock-work melted. Steel pivots of wheels magnetised. Silver face of regulator blackened. Door of clock casing	Harr. 80.

SOME INCIDENTS OF THE ACTION OF LIGHTNING.—(*Continued.*)

Place.	Object.	Particulars.	Lightning rods present.	Metals present.	Results.
				ing cramps to masonry. (6.) Lead floor to cupola and lead joints to stones. (7.) Framework of iron and wood resting on floor. (8.) Iron spindles of clock faces in dial-room. (9.) Bells weighing 5 to 31 cwt. (10.) Massive iron clock frame. (11.) Clock works. (12.) Vertical iron rod, ¾ in. diameter, 46 ft. long, in lengths united by brass screws, connecting dial hands with clock-works. (13.) Lead-covered roof of church.	burst open. Clock not stopped. Clock-room floor left "as if blown up by gun-powder." Glass of iron window-frame shattered. Portions of lead joints of stones and of lead roof slightly fused. Damage everywhere occurred at points where metal ceased. Discharge reached earth by means of "large masses of metal and pipes connecting the roof with the ground."

| 52 | 24 Apl., 1842. | London. | Brixton Church. | Has a dome of massive masonry, on columns 12 ft. high. Clockwork roof below, then belfry, on level with church roof. | ... | ... | (14.) Rain-pipes of iron or lead from roof to ground. (1.) Copper cross 4½ ft. high and 5 ft. square. (2.) Horizontal iron bars supporting it. (3.) Dome joints soldered with lead. (4.) Lead roof to clock-room. (5.) Iron connecting rods to dial plates, ½ in. diameter. (6.) Copper dial plate, 4 ft. diameter, and leaf-gold figures. (7.) Iron wire 1/16 in. diameter, 30 feet long, between clock-work and bells, of several pieces looped together. | Dome masonry rent open. Base of column, at junction with leaden roof of church, shattered. Wire between clock-work and bells knocked to pieces and partly dissipated. Discharge made earth by means of rain-pipes. | Harr. 86. |

SOME INCIDENTS OF THE ACTION OF LIGHTNING.—(*Continued.*)

Place.	Object.	Particulars.	Lightning rods present.	Metals present.	Results.	Authority.
Ham, France.	Abbey of Notre Dame.	..	..	(8.) Bells. (9.) Iron cramp on belfry floor. (10.) Lead roof of church. (11.) Rain-pipes (iron or lead) to ground.	Struck 3 times in 20 minutes. The whole abbey burnt to the ground.	Harr. 36.
Edinburgh.	The "Dasses;" crag of rock in a valley, surrounded by high land and by overlooking prominent rocks.	Of whinstone, with a flat top covered with sod.	..	..	Sod-covered top struck and several tons of rock detached. A hole made in sod 2 ft. 6 in. long by 1 ft. 6 in. broad, and a furrow ploughed therefrom in the sod for a distance of 19 yards, 5 in. wide at hole, and diminishing in width as it receded. Compass needle deviated from N. to E.S.E. when afterwards placed in hole.	Tel. 1/8/80.

98

No.	Date	Place	Building					Remarks	Reference
55	July, 1880 (5.30 a.m.)	Near Battle, Sussex.	Normanhurst Court. (Built of stone with tiled roof, on high land, in exposed situation, not far from English Channel.)	Large country house, with a tower and spire 120 ft. high at one corner; At another, a chimney-shaft 126 ft. distant, with a coping of 8 pinnacles.	Copper wire bands 1 in. broad on tower and chimney. Terminal rod at chimney of copper tube 6 ft. long, with iron point 12 in. long, rising 3 ft. above coping, united by joints. Rod insulated from chimney-shaft and terminating in dry earth.	..	..	Rod at chimney-shaft struck and bent, and lead composition on iron point melted. Some wires near terminal partly fused, and damage done to wires 25 ft. lower down. Three pinnacles destroyed.	Tel. 1/10/80. (Mr. R. Anderson.)
56	8 Sept., 1880.	Paris.	The Sorbonne. (University buildings.)	..	Six stems connected by an iron bar making a circuit over whole roof, joined to an iron rod less than 1½ inch square, leading into a pit at a great distance from buildings.	..	..	Two rods struck. ("Only erected "within the last few months, and never before, as far as is known, had any thunderbolt struck the venerable abode of the French University.")	Tel. 1/10/80.
57	1769.	Sienna, Italy.	Cathedral.	Furnished this year with a rod.	..	..	..	Has not suffered since rod was fixed, though had repeatedly been damaged before.	Harr. 96. Ar. 262.

III. — SOME INCIDENTS OF THE ACTION OF LIGHTNING.—(Continued.)

No.	Time.	Place.	Object.	Particulars.	Lightning rods present.	Metals present.	Results.	Authority.
58	June, 1839.	Paris.	Hotel des Invalides.	..	Rod of 20 iron wires twisted as a small rope.	Lead around lantern.	Rod broken into pieces and scattered. Lead torn up and scattered.	Harr. 104.
59	Dec., 1824.	Plymouth.	Charles Church.	With steeple.	Small rod of linked brass.	..	Rod torn in pieces. Church and tower not damaged. [*Note by Author.*—Granite spire on tower of moderate height, with massive spindle, ball, and vane. Pavement at base of tower. Church in middle of town. Site low, not exposed.]	Harr. 104, 160.
60	Nov., 1790.	Portsmouth harbour.	H.M. Ship *Elephant*.	..	..	Iron hoops and mouldings on main-mast.	Hoops and mouldings burst open and broken to pieces, but none melted.	Harr. 103.
61	Dec., 1838.	..	H.M. Ship *Rodney*.	..	..	Copper funnel 16 in. long, 10 in. diameter, and less than ¼ in. thick, in main top-gallant rigging. Iron hoops 5½ in. wide and ¾ in. thick on main-mast.	Main top-gallant mast dispersed in small chips. Main top-sail set on fire. Piece 10 ft. long torn out of main top-mast. Main-mast had 13 hoops burst and was damaged for a length of 53 ft.	Harr. 110.

10 June, 1764.	London.	St. Bride's Church.	Stone spire formed of an obelisk resting on four successive stories of different orders of architecture, below which is belfry.	(1.) Ball, cross, and vane of gilt copper. (2.) Iron spindle of vane 2 in. square, 20 ft. long, let in with lead. (3.) Iron collars, soldered with lead, connecting upper courses of obelisk. (4.) Horizontal iron bars connecting piers in each story with iron chain tie bars. (5.) Iron cramps to masonry. (6.) Iron bars supporting window heads.	Lightning made successive leaps between the metals, rending the masonry where it entered and where it left it. Gilding at top of cross discoloured. Last trace at west belfry window, where it appears to have been led to earth. A stone of 70 lbs. thrown to 50 yards. An iron clamp ½ inch thick and 2 ft. long broken in two, and one part bent at an angle of 45°.	Harr. 89, 102. Ar. 142.
Sept., 1833.	Indian Ocean.	H.M. Ship *Hyacinth*.	...	Iron chains below main and fore top-masts, ¼ in. diameter, and 50 ft. long. Copper pipe in hull.	Two successive discharges. Main and fore top-gallant and top-masts shivered. Carried away below top-masts by chains. Passed into sea by copper pipe.	Harr. 110.

III. SOME INCIDENTS OF THE ACTION OF LIGHTNING.—(Continued.)

No.	Time.	Place.	Object.	Particulars.	Lightning rods present.	Metals present.	Results.	Authority
64	Summer, 1760.	Philadelphia, North America.	House.	Mr. West's.	Iron rod ½ in. diameter 9 ft. above chimneys, ending in an iron stake in ground (dry). Brass point 10 in. long and ¼ in. diameter.	...	Rod visibly struck. Brass point fused for 3 inches in depth. Flame appeared at base of rod. House uninjured. (Rod erected by Franklin.)	Harr. 111. And. 30
65	26 Jan., 1858.	Rio Janeiro Harbour, South America.	H. M. Ship *Dublin*.	...	Rod of copper links ¼ in. diameter and 10 in. long.	...	Rod struck. Links melted in places.	Harr. 111.
66	17 June, 1774.	Tenterden.	House.	...	Iron bar ¾ in. square.	...	Rod struck. No injury.	Harr. 111.
67	June, 1772.	Steeple Ashton, Wilts.	Vicarage.	...	...	Iron bell-wires in parlour and hall.	Bell-wires dispersed except where doubled or twisted.	Harr. 112.
68	June, 1828.	Kingsbridge, Devonshire.	Church.	With tower and spire.	...	Iron spindle surmounting spire 1 inch diameter, 7 ft. long.	Tower shattered. Spindle uninjured.	Harr. 112.
69	Aug., 1822.	Rio de la Plata, South America.	H.M. Ship *Beagle*.	...	Copper plates $\frac{3}{16}$ inch thick and 1 to 5 in. broad fixed in masts leading from copper spindles ¼ in. diameter 2 ft. long and ending in hull metals and sea.	...	Ship struck. No damage.	Harr. 112.

Coast of Central America.	H.M. Ship *Actæon*.	..	Copper plates, &c., as No. 69.	Ship struck. No damage.	Harr. 112.
Calcutta Harbour.	H.M. Ship *Endymion*.	Fore-mast 50 ft. from main-mast.	Chain rod on main-mast 150 ft. high.	Fore-mast struck. Top-gallant and top-masts shivered.	Harr. 118.
Corfu Harbour.	H.M. Ship *Etna*.	..	Chain rod on main-mast.	Several discharges. Rod struck, also ship near bows, and some persons there knocked down.	Harr. 118. Ar. 264.
Heckingham.	Poor House.	Had eight chimneys.	Pointed rod on each chimney.	A rod struck, also a corner of house 70 ft. distant from nearest rod, also a gate in front of house.	Harr. 119.
Edinburgh.	Melville Monument.	..	Pointed rod through statue, thence iron chain to ground.	Slight damage. (Chain found to have been lifted out of ground by some mischievous person).	Harr. 125.
Coast of Africa, Atlantic Ocean.	H.M. Ship *Leven*.	..	Had rod, but moveable lower extremity was in its box at the time.	Channels of ship destroyed.	Harr. 139.
Corfu Channel.	H.M. Ship *Madagascar*.	When coming to anchor.	..	Struck 5 times in 2 hours.	Harr. 138. Ar. 264.
Charlestown, North America.	St. Michael's Church.	..	Rod erected this year.	Frequently struck and damaged before 1760, but never since erection of rod.	Harr. 160. Ar. 261.

III. Some Incidents of the Action of Lightning.—(*Continued.*)

No.	Time.	Place.	Object.	Particulars.	Lightning rods present.	Metals present.	Results.	Authority.
78	1765.	New York.	Dutch Church.	With clock tower.	Rod erected this year.	Clock and bell.	Rod struck, but no damage done.	Harr. 160. Ar. 261.
79	1783.	Carinthia, Hungary.	Chapel of Count Orsini's Castle.	With bell tower. On elevated site.	With rod.	Bell (presumed).	Rod struck, but no damage done.	Harr. 161. Ar. 260.
80	1772.	Turin.	Valentino Palace.	..	Rod erected this year.	..	Frequently damaged before 1772, but not struck after erection of rod.	Harr. 161. Ar. 261.
81	1766.	Venice.	St. Mark's Cathedral.	Tower and spire 340 ft. high.	Rod erected this year.	Wooden figure of angel covered with copper on summit, also detached pieces of iron in tower.	Struck and damaged in years 1388, 1417, 1489, 1548, 1565, 1653, 1745, 1761, and 1762. In 1745 the tower was rent in 37 places. Not suffered since erection of rod.	Harr. 162. Ar. 262.
82	May, 1782.	Glogau, Silesia, Prussia.	Powder magazine.	A sentry guarding it.	Pointed rod ending in a well of water.	..	Magazine struck. Sentry received a shock.	Harr. 163.
83	20 Sept. 1880.	Bishop Stortford.	Parish Church.	..	..	Gas-pipes.	Vestry roof penetrated, woodwork charred, gas-pipe melted.	Stand. 21/9/80.

84	Aug., 1769.	Brescia, Italy.	St. Nazaire's Church.	With tower. Vaults used as a powder magazine.	..	..	Tower and vaults struck. 207,000 lbs. of powder exploded. One-sixth of Brescia destroyed. 3,000 persons killed.	Harr. 164. (Quoting Arago.)
85	Sept., 1880.	Hendon, near London.	Lake.	At the "Welsh Harp."	..	..	Struck. A ball of fire seen. 100 fishes found afterwards floating dead on the surface.	Tel. 1/10/80.
86	10 Oct., 1878.	London.	Furniture warehouse.	At the back of Victoria Railway Station. Brick building 110 ft. by 80 ft. Flat roof, slate covered tower at one corner, 20 ft. above roof.	Copper band ¾ inch wide by ⅛ inch thick with terminal tube ⅜ inch diameter, and 3 points 6 in. long, all copper, attached to and rising 3 ft. above iron cresting. Gunmetal holdfasts.	(1.) Lead covering to flat roof. (2.) Iron cresting 4 feet high on roof of tower. (3.) Metal rainpipes from lead roof to ground.	Rod struck. Tower considerably damaged. Iron cresting shattered. Rod bent, two points driven out, and third twisted. (Earth connections found afterwards to terminate in concrete.)	Tel. 1/10/80.
87	June, 1807.	Luxemburg.	Powder magazine.	On a solid rock.	..	..	Blown up. 28,000 lbs. of powder exploded. Lower part of town ruined.	Harr. 164. (Quoting Arago.)

III. SOME INCIDENTS OF THE ACTION OF LIGHTNING.—(*Continued.*)

No.	Time.	Place.	Object.	Particulars.	Lightning rods present.	Metals present.	Results.	Authority.
88	25 June, 1880.	Sutton, near Birmingham.	Mess tent and band tent in a Volunteers' encampment.	..	..	Cooking stove. Band instruments.	Centre pole of mess tent shivered to pieces and 3 persons out of 100 sheltering there injured. A cook knocked down while at work. Band tent struck, and several bandsmen practising in it injured.	W.M.N. about 27/6/80.
89	10 June, 1880.	Skelton.	St. Mark's Church.	..	Copper rope ⅜ inch diameter. Led into "earth" 9 inches deep in bricks, lime, &c. 15 ft. from church.	..	Rod struck and slightly bent.	Tel. 1/10/80. (Mr. R. Anderson.)
90	About 18 June, 1880.	Near Nottingham.	Arthur Derrick, a boy 8 years old.	In a field	..	..	Killed.	W.M.N about 19/6/80.
91	26 June, 1880.	South Lambeth, London.	All Saints' Church.	Stone cross 4 ft. high on apex of west gable. Similar one 92 ft. distant at east end of nave.	Copper rope ⅜ inch diameter, with copper tube attached to end, rising 5 ft. above west cross. Led 2 inches deep into loose rubbish.	..	Rod struck. Roof of north aisle injured and east cross thrown down.	Tel. 1/10/80. (Mr. R. Anderson.)

No.	Date	Place	Object	Situation	Protection	Result
92	June, 1770.	Philadelphia, Nth. America.	3 houses and a ship.	..	Pointed rod on one house.	Rod struck and point melted, but no damage to house. Other houses and ships suffered considerably.
93	24 June, 1880.	Acrise, Kent.	Oak tree.	..	..	Slip of bark 3 inches wide taken off and a groove made $\frac{1}{8}$ inch. deep in body of tree.
94	..	Jerusalem.	The Temple.	On an elevated exposed position subject to frequent and heavy storms.	..	Covered inside and outside with metal plates. Roof thickly gilded and covered with pointed iron spikes. Metal rain-pipes from roof to ground. Never struck.
95	..	Geneva.	Cathedral.	Most prominent and elevated site.	..	Central tower covered with tinned iron plate connected to ground by metal pipes. For two centuries not struck, whilst bell tower of St. Gervais Church on a much lower site was frequently struck and damaged.

III. Some Incidents of the Action of Lightning.—(*Continued.*)

No.	Time.	Place.	Object.	Particulars.	Lightning rods present.	Metals present.	Results.	Authority.
96	25 May, 1841.	Plymouth.	Chimney-shaft of bakehouse of Royal William Victualling yard.	Round granite shaft 120 feet high.	..	At 60 ft. below summit is extensive copper roofing of bakehouse, connected to ground by metal pipes.	Chimney rent for 60 feet as far as copper roof, where damage ceased. [*Note by Author.*—The track of the lightning is still visible, where the shaft has been repaired by iron cramps. Shaft is near harbour, and ground is only a few feet above H.W.M.]	Harr. 168 and 180.
97	Summer, 1774.	London.	St. Peter's Church.	With brick tower and spire.	..	Large copper-gilt key. Lead covering to spire. Lead roof to church. Thence metal pipes to ground.	Between lead spire covering and lead roof the tower was much rent. No other damage.	Harr. 168.
98	About 18 June, 1880.	Midland Railway.	A plate-layer named Henry.	At work on the railway with 2 other men.	..	..	Struck, and died on way to hospital. The two other men injured.	W. M. N. about 19/6/80.
99	1830.	Coast of Africa, Atlantic Ocean.	H.M. Ship *Dryad*.	..	Had rods	..	Struck, but received no damage.	Harr. 171.
100	1832.	Rio Janeiro.	H.M. Ship *Druid*.	..	Had rods	..	Struck, but received no damage.	Harr. 171.

Date	Location	Structure				Reference
1831.	The Tagus.	H.M. Ship *Asia*.	..	Had rods ..	Struck, but received no damage.	Harr. 171.
1842.	Sheerness Harbour.	H.M. Ship *Tallat*.	..	Had rods ..	Struck, but received no damage.	Harr. 171.
..	New Chester, Nrth. America.	House.	With a chimney stack at each end. Shingled roof. On a rocky base at foot of a steep ascent, a few yards from a mill dam.	A rod on each stack passing down wall directly to rock.	Near west rod a copper water-gutter and rain-pipe passing down wall to within 4 ft. of a cistern on ground.	West rod struck and its end fused. Ground at its foot torn up. Roof shingle torn up between west chimney stack and copper gutter. Damage also done lower down, between gutter and rod.	Harr. 217.
May, 1835.	..	H.M. Ship *Racer*.	..	A chain rod on main-mast.	..	Rod struck—sparks being seen on it. Also fore top-gallant mast struck, distant 40 ft.	Harr. 221.
May, 1837.	Chowringee, India.	Two houses.	66 ft. apart.	A rod on one house.	..	Both houses struck. The one with rod not hurt, the other damaged.	Harr. 221.
29 Jan. 1836 (night).	Black Rock, Cork, Ireland.	St. Michael's Church.	With limestone spire.	..	Spire "strengthened by iron cramps and bars in the usual way, within and without."	Top of spire thrown down, and the whole of the windward side of the spire rent open.	Harr. 91.

III. Some Incidents of the Action of Lightning.—(*Continued.*)

No.	Time.	Place.	Object.	Particulars.	Lightning rods present.	Metals present.	Results.	Authority.
107	Oct., 1836.	Doncaster.	Christ Church.	With a magnificent spire.	..	Had a vane surmounted by a glass ball.	Nearly half of spire demolished.	Harr. pref. and 130.
108	14 Aug. 1779.	Genoa.	Church of Notre Dame, de la Garde.	With a bell tower. On one of the highest hills of the neighbourhood.	Stout iron rod with gilt copper point partially inside tower, and projecting 3 ft.	..	Rod struck. Point split open. Porch at some distance off also struck and church near it damaged.	Harr. 209.
109	25 Mar., 1840.	Vourla Bay, Mediterranean.	H. . Ship *Powerful*.	At anchor.	..	..	Struck, and some spars shivered. (H.M. Ship *Asia*, a short distance off, with a rod, did not suffer.)	Harr. 181.
110	14 July, 1880.	Pennsylvania, Nth. America.	Iron oil tanks.	..	With rods	..	Struck, and 200,000 barrels of oil consumed.	W.W.N. 7/8/80. Tel. 15/8/80.

110

111

3 Aug., 1879.	Wells-next-the-Sea, Norfolk.	St. Nicholas Church.	With lofty square embattled tower of dressed flints, at west end (without pinnacles). Vestry near chancel at east end. Three stained glass windows at east end and chancel. Roof woodwork of chestnut.	..	Eight bells and clock in tower. (Stained windows had probably usual wire coverings.)	At east face of tower a large portion of stonework driven out. Discharge through roof along nave to vestry. Church and vestry simultaneously set on fire, and in two hours was a mass of ruin. The three stained windows destroyed. Seven of the bells were melted.	I.L.N. 30/8/79.
31 May, 1875 (night).	Concordia, Argentine Republic.	Brazilian Vice Consul's house.	One storied. Flat tiled roof with parapet wall. Flagstaff 18 feet high over front door. Front wall plastered. Floor tiled at front door.	..	Iron staples holding flagstaff to parapet. Iron shield with coat of arms hung over front door. Bolt at top and bottom of door, also lock and hinges.	A man and boy in an adjacent room thrown down. Flagstaff shattered longitudinally into several pieces. One piece 10 feet long and 2 inches broad thrown on roof of an adjacent house 50 yards away. Bricks and tiles of roof dislodged in two	Tel. 15/9/75. (Mr. J. H. Blomfield, of Concordia.)

III.—SOME INCIDENTS OF THE ACTION OF LIGHTNING.—(*Continued.*)

No.	Time.	Place.	Object.	Particulars.	Lightning rods present.	Metals present.	Results.	Authority
							places each 2 feet square, about 6 feet from heel of flagstaff. Plaster of front wall cracked. Opening made in wall from staple at heel of flagstaff to iron shield. Door frame split and charred, and paint blackened between shield, bolt, lock, and hinges. Tiles torn up at foot of lower bolt.	
113	1875.	Whitewater, Michigan, U.S.	Mr. A Castle, a farmer.	Sitting in a barn, where, with a team of horses, had taken shelter from storm.	..	..	Both horses killed, and farmer prostrated insensible. On reviving, felt an intense burning pain—air and sky seemed ablaze. Legs paralyzed for eight or ten hours. A broad irregular strip, from which skin peeled off, between calf of right leg and shoulders.	Tel. 15/10/75.

114	31 July, 1877.	Barrackpore, India.	Royal Artillery Barracks.	Two storied, with verandahs and flat cemented roofs.	Iron rod 1½ inches diameter at base, projecting 5 ft. above roof parapet, with a good "earth" of iron chains.	Iron girder of verandah roof running 2 in. below horizontal part of rod.	Rod struck. Small hole pierced through cement of verandah roof between rod and girder, 9 inches long and 2 inches deep. (Junctions of vertical and horizontal parts of rod afterwards found defective.)	Mann, J.S.A. 1878, 328.
115	About 1878.	Natal, South Africa.	A Dutch farmer named Buys.	Entering his house, which was one storied.	Two galvanised iron rods ½ inch diameter, with brush points, but with lower parts broken off, and no earth made.	Galvanised iron roof.	House struck. Farmer killed. Two children in the house hurt.	Mann, J.S.A. 1878, 330.
116	About 1878.	Inchanga Hill, Natal, South Africa.	16 natives and 5 oxen.	On the high road.	..	..	Killed at one discharge.	Mann, J.S.A. 1878, 330.
117	25 Feb., 1875.	Paignton, Torbay.	Flagstaff before a house.	50 feet high. On a bold headland projecting into Torbay. (Just after a burst of hail.)	..	Metal vane on top. Stayed at 25 ft. from ground by 4 galvanised iron wire ropes, ending at 1 ft. from ground in ½ in. iron chains,	Flagstaff above wire stays split into fragments. Small craters formed in the ground where the chains entered. Soil loosened for about 12 inches deep. Ground affected	Mann, J.S.A. 1875, 518.

III. SOME INCIDENTS OF THE ACTION OF LIGHTNING.—(*Continued.*)

No.	Time.	Place.	Object.	Particulars.	Lightning rods present.	Metals present.	Results.	Authority.
118	16 Oct., 1868.	Nottingham.	All Saints' Church.	Spire 150 feet high.	Rod of 7/16 inch copper rope, with end turned round a stone just below the ground.	Gas-pipe inside church against the wall outside which the rod ran, 6 ft. above ground.	Rod struck. Wall 4 ft. 6 inches thick between gas-pipe and rod pierced.	Mann, J.S.A. 1875, 518.
						corroded in places to ⅛ inch. Chains anchored in dry red sandstone conglomerate.	about 30 ft. off. Three chains broken, 20 links being snapped across. Mrs. Pidgeon standing at the door of the house, 10 ft. from a chain, felled to the ground. Mr. Pidgeon, standing 10 ft. from a chain, received a shock.	
119	1865.	England.	House.	Near that of Mr. G. J. Symons.	..	Water-pipes, gutters, gas-pipes, bell-wires.	Chimney and fire-place struck, and coals scattered, by one flash. Gas lighted and bell wires dissipated by another, which also traversed water-pipes and gutters. Another flash disappeared.	Sym. S.T.E. 358.

115

120	About 1872.	England.	Two houses and an adjacent beech tree.	One house was Mr. Latimer Clark's.	Rod on one house.	..	Rod struck by one stroke. The other house by a second. The tree by a third, and a rake leaning against it was split into shreds.	Lat. Clark, S.T.E. 373.
121	1872.	Spain.	The Escurial Palace.	..	..	..	Struck, and partially destroyed by fire arising from lightning. (Has three times before been on fire from the same cause; but never yet has been protected by a rod.)	Preece, S.T.E. 336.
122	Aug., 1880.	Castle Gregory, Tralee, Ireland.	A man.	Sheltering under a tree.	..	..	Killed.	W.W.N. 21/8/80.
123	29 May, 1880.	Cole Harbour, Halifax, N.S.	A diver.	Under water.	..	..	Rendered insensible. Air pump above struck.	Tel. 15/7/80.
124	About 1872.	.. ..	Tree.	..	..	..	Struck. Bark of lower 40 ft. stripped off. Above the 40 ft. line the track of the flash was distributed among the branches.	Grav. S.T.E. 413.

III. SOME INCIDENTS OF THE ACTION OF LIGHTNING.—(*Continued.*)

No.	Time.	Place.	Object.	Particulars.	Lightning rods present.	Metals present.	Results.	Authority.
125	4 Sept. 1880 (night).	Coast off the Tyne.	Schooner *Glanogwen.*	Of Beaumaris, from Carnarvon to Dundee, with slates.	..	..	Main-mast shattered all to pieces. [*Note by Author.*— This is the only record of a casualty at sea from lightning, in 1880, that has been obtainable.]	Stand. 7/9/80.
126	26 June, 1880 (3 P.M.)	London.	House, 180, Oakley St., Chelsea.	..	..	..	Kitchen chimney stack shattered. A lump of fused brick, mortar, soot and cinder fell in kitchen grate.	S.M.M. Aug. 80.
127	Dec., 1854.	Paris.	The Louvre.	..	Rods erected in 1782 by Le Roy.	..	Trifling damage. Great alarm. (Storm swept along the banks of the Seine.)	And. 80.
128	30 Apl., 1822 (even.)	Rostall, Franconia, Bavaria.	Church.	With steeple 156 feet high. On brow of a hill on dry sandy soil.	Brass wire rope over 1 inch diameter (designed by Von Yelin.)	Clock in steeple and other metals.	Rod struck, and point melted. Flash divided, part to clock and other metals and part down rod. Clock thrown from its place and part of lower wall thrown down.	And. 165.

Date	Location	Building	Description of building	Description of rod	Damage	Reference
6 Aug., 1878 (4.30 P.M.)	Victoria colliery, Bruntcliffe, Yorkshire.	Powder store.	Oblong brick building, 9 feet × 5 feet × 6 feet high internally. Walls three bricks thick. Brick arched roof. Contained 2,000 lbs.	Copper wire rope at one end, $\frac{7}{8}$ inch diameter, fixed to a pole about 2 in. from building with glass insulators. Point projected 13 ft. above building, with a brush of 4 points. Rod went 1 ft. deep into ground for 13 ft., and ended in a drain.	Store blown up. Two little girls 320 yards distant wounded by broken *débris*, also adjacent buildings damaged by same cause.	Rep. Expl. 17/9/78.
Aug., 1879.	Cromer, Norfolk.	Church.	Fine Gothic stone building, with tower 159 ft. high, and pinnacles.	Good rod on one pinnacle of $\frac{5}{8}$ inch copper. Earth connection in thoroughly good order.	...	A pinnacle 27 ft. 6 in. distant from the one carrying rod was struck and damaged.	And. 147.
May, 1879.	Laughton-en-le-Morthen.	Church.	With steeple 175 ft. high.	Copper tube, $\frac{1}{4}$ inch thick and $\frac{3}{4}$ inch external diameter, in short lengths, joined by screws and coupling-pieces, distant 6 ft. 6 in. from lead roof. No metallic contact between pieces, which were much corroded. Fastened to vane and	Vane on steeple. Lead-covered roof attached to steeple. Cast-iron down pipes thence to earth.	Rod thrown down and broken in two pieces. Buttress between rod and lead roof pierced, and about 2 cart-loads of stone dislodged.	And. 153.

III. SOME INCIDENTS OF THE ACTION OF LIGHTNING.—(Continued.)

No.	Time.	Place.	Object.	Particulars.	Lightning rods present.	Metals present.	Results.	Authority.
132	1 Aug., 1846.	Leicester, England.	St. George's Church.	With tower, pinnacles, and steeple.	kept from building by 21 insulators. Rod buried 6 to 18 inches in dry, loose, rubbish, for a length of 5 feet.	(1.) Spindle, (2.) Gilt vane. (3.) Iron vane supports. (4.) Iron cramps to sandstone blocks of steeple. (5.) Copper bolts through pinnacle stones. (6.) Cast-iron pipe from tower to church roof. (7.) Clock works and dial faces. (8.) Lead roofs and gutters of church and side lobbies. (9.) Iron pipes to earth from roofs.	Vane struck. Cramps magnetised. Sandstone blocks torn and hurled aside. Pinnacle struck. Clock works struck. Explosions occurred at overlappings of metals on lead roofs.	And. 177. Quoting Mr. C. Tomlinson. F.R.S.

133	27 Sept. 1875.	Oxford.	Merton College Chapel.	Tower, with pinnacles 25 feet above roof of tower.	..	Vane on pinnacles. Lead roof to tower.	Pinnacle severed from summit to base. Vane slightly fused and found embedded in lead roof.	And. 182.
134	8 June, 1879 (3.30 P.M.)	Wrexham.	St. Mark's Church.	With spire.	Copper rod of small calibre.	..	Spire struck. Some stonework displaced. Six persons in a room at base of tower (forming a Sunday class) injured by burning.	And. 186.
135	18 Dec., 1725.	Winterthur, Germany.	Church.	With tower.	..	Metals forming an "accidental conductor."	"Accidental conductor" melted.	And. 186.
136	21 July, 1745.	Bologna, Italy.	Monastery.	With tower.	..	As No. 135.	"Accidental conductor" melted.	And. 186.
137	13 July, 1880.	Manchester.	House in Bridge Street.	Four storied. Abutting on the river Irwell.	..	..	House struck at gable facing the river, and collapsed, a heap of ruins, into the river, some large stones being shattered at its base. William Way (a pedler) and Annie Jones (aged 17) were buried in the ruins and killed. Seven others more or less injured.	W.W.N. 17/7/80.
138	31 May, 1748.	Witzendorf, Germany.	Church.	With tower	..	As No. 135.	"Accidental conductor" melted and shattered. Roof torn off and shattered.	And. 186.

III. SOME INCIDENTS OF THE ACTION OF LIGHTNING.—(*Continued.*)

No.	Time.	Place.	Object.	Particulars.	Lightning rods present.	Metals present.	Results.	Authority.
139	6 June, 1751.	South Molton, Devonshire.	Church.	With tower.	..	As No. 135.	"Accidental conductor" shattered.	And. 186.
140	16 June, 1754.	Newbury.	Church.	With belfry.	..	As No. 135.	"Accidental conductor" melted.	And. 186.
141	13 July, 1880.	Oldham.	Two mills.	..	..	..	Struck and set on fire, but soon extinguished.	W.W.N. 17/7/80.
142	16 July, 1760.	Altona, Denmark.	Church.	With spire.	..	Copper covering on spire and metals forming "accidental conductor."	Copper covering struck. "Accidental conductor" melted.	And. 186.
143	6 Aug., 1767.	Hamburg.	Nicholas tower.	..	..	As No. 135.	"Accidental conductor" partly melted.	And. 186.
144	2 Feb., 1771.	Kiel, Denmark.	Nicholas Church.	..	..	As No. 135.	Traces left on "accidental conductor."	And. 186.
145	March, 1772.	London.	St. Paul's Cathedral.	..	..	As No. 135.	Traces left on "accidental conductor."	And. 186.
146	25 June, 1880.	Alton, Hants, England.	Church.	With shingled spire.	..	..	Shingle covering of spire removed in a zigzag line from top to bottom and some beams in spire splintered.	W.M.N. About 27/6/80.

147	Feb., 1823.	Shaugh, near Plymouth.	Church.	With tower.	Iron rod erected 2 years before. In rusty state.	Tower struck and damaged. [*Note by Author.*—Granite tower with pinnacles, moderate height, very elevated ground (borders of Dartmoor), much exposed. Rocks cropping up in adjacent ground, but plenty of grassy soil between them.]	And. 186.
148	26 June, 1880.	York Road, Wandsworth, London.	Two policemen.	On duty.	...	Struck and rendered insensible.	English paper, about 28/6/80.
149	26 June, 1880.	Trent.	Two labourers named Brooks and Brown.	Sheltering under a tree.	...	Brooks was killed. Brown received a shock.	English paper, about 28/6/80.
150	June, 1854.	Ealing, near London.	Church.	...	Had a "common conductor."	Rod struck and fused. Church slightly damaged.	And. 186.
151	26 June, 1880.	Near Wilton Towers, Wilton-le-Wear, Durham.	Mr. Henry Smith Stobart, J.P., of Wilton Towers.	While walking on a railway with Mr. T. Elliott.	...	Killed. (No mention of occurrence in *Standard* except advertisement of death.)	Stand. and W.M.N. June, 1880.
152	26 June, 1880.	Bradninch, Devonshire.	Poplar tree.	Near Railway Inn in Hele Square.	...	Tree split, discharge passed through passage of inn.	W.W.G. about 1/7/80.

SOME INCIDENTS OF THE ACTION OF LIGHTNING.—(*Continued.*)

No.	Time.	Place.	Object.	Particulars.	Lightning rods present.	Metals present.	Results.	Authority
153	July, 1854.	Ashbury.	Church.		Had a "common conductor."		Rod struck and fused. Church slightly damaged.	And. 186.
154	Aug., 1857.	St. Luke's, London.	Gasholder.	Chartered Gas Company's Works.			Struck, and gas ignited.	And. 186.
155	July, 1779.	Genoa.	St. Mary's Church.	Very elevated position. A pool of water on one side of church.	A rod of "most approved design," erected in November, 1778, making earth in hard rock on other side of church.	Iron cramps connecting hewn stones of structure. Much metal inside and outside church.	Church struck. Discharge demolished masonry and took the same path as it had frequently done before the erection of rod, *i.e.* along cramps and metals, and thence to pool of water.	And. 201.
156	2 Oct., 1872.	Alatri, Italy.	Cathedral.	With belfry. On high ground much exposed. Solid calcareous rock a short depth below ground.	Two separate copper rods on belfry and choir respectively. Copper earth connections 13 feet long, with points in them, and copper wire twisted among the points and laid in carbon.	Waterwork pipes supplying Alatri and Ferrentino towns, passing about 11 yards from earth connections. Lead slab to tank, and spouts to fountain, at Alatri.	A rod struck and 1¼ in. of point melted. Earth excavated for 28 in. deep between earth connections and water-pipes. Ferrentino pipes broken. Slab of tank damaged. Trace of course on spouts of fountain. No injury to church. (Rods erected in 1864, and church struck at least 4 times subsequently, but without damage.)	And. 203. (Quoting Father Secchi.)

No.	Date	Place	Building		Conductor	Contents	Effects	Authority
157	15 March, 1876.	Clevedon, Somersetshire.	Christ Church.	With tower containing a flagstaff 100 ft. high, and 4 pinnacles 90 ft. high. A house 100 yards off supplied by gas-pipes. Soil of neighbourhood, mountain limestone, very dry. No rain accompanied flash.	Five copper wire ropes of ¼-in. diameter, one on each pinnacle, and one on flagstaff, passing down inside tower and uniting near clock, passing outside lower down and encased for lowest 12 ft. in a pipe, and passing below ground into a dry freestone channel for about 12 ft., and then into a rain-water drain.	Gas meter and pipes inside church against wall 3 ft. thick, outside which rod passed, with pipes thence to house, where passed 1 in. from water-pipes.	Rod struck. Wall between rod and gas meter pierced. Shock felt in house. Discharge supposed to have passed there to water-pipes and thence to have made earth, since opposite them a hole ⅝-in. diameter found in gas-pipe, which was made of composition metal.	123 And. 208. (Quoting Mr. Eustace Butter, in S.T.E.)
158	..	Lyons, France.	House.	Belonging to a banker.	Had "a good conductor."	A large iron safe in the house.	Safe struck. Some gold melted and bank notes burnt. ("Not many years ago.")	And. 220.
159	Summer, 1871.	Halifax, Nova Scotia.	Old Provincial Buildings.	..	..	Gas-pipes partly of composition metal, and gas meter in basement.	Gas-pipe fused and gas lit. Building nearly set on fire.	W.O. 1875.
160	28 July, 1857.	Jamaica.	Compton Lodge.	..	Iron rod 10 ft. distant from S.E. corner of building.	..	S.E. angle of building shattered in pieces. Rod untouched. Escape of family miraculous.	W.O. 1858. App.

III. SOME INCIDENTS OF THE ACTION OF LIGHTNING.—(*Continued.*)

No.	Time.	Place.	Object.	Particulars.	Lightning rods present.	Metals present.	Results.	Authority.
161	Aug., 1875 (afternoon).	Haydon Bridge, Northumberland.	Mr. Nicholas Woodman, a farmer.	Riding along a road in advance of a gig.	..	Watch	Mr. Woodman and horse killed, also the horse in gig. Left side of Mr. Woodman terribly burnt, and his coat, waistcoat, and watch consumed.	Tel. 1875. 180.
162	Whit-Monday, 1874.	Gardens of White Hart Tavern, Hackney Wick, London.	Mr. William Roberts, a greengrocer.	Sheltering under an elm tree, with 7 others.	..	..	Mr. Roberts killed. Hair and whiskers singed, hat torn, mark down left side as far as ankle. Others received a shock.	Times, 28/5/74.
163	25 May, 1874 (afternoon).	Field's End Farm, Berkhampstead.	Farm buildings.	Mr. John Ginger's.	..	..	Buildings and produce destroyed. Several thousand pounds' worth of damage.	Times, 28/5/74.
164	25 May, 1874 (noon).	East Malling.	William James, a servant of Mr. Arkale, and a boy named Hales.	Had been looking on at a cricket match, and were sheltering during a storm under a tree.	..	..	Killed.	Times, 28/5/74.

165	About 24 June, 1874 (3 P.M.)	Flatfield farm, near Cupar, Angus, N.B.	Alexander Airth, the grieve; Mrs. Wallace, wife of shepherd, and 7 others.	Hoeing turnips when storm came on, then took shelter under a tree close by. With spire.	..	Mr. Airth killed. Trousers and shirt on left side much torn. Mrs. Wallace injured on legs. The others received a shock.	Times, 26/5/74.
166	About 24 June, 1874 (afternoon).	Braco, near Cupar, Angus, N.B.	Free Church.	With spire.	..	Spire twisted. Stones round its base loosened. Roof of church shattered.	Times, 26/5/74.
167	About 24 June, 1874 (afternoon).	Lossiemouth, Morayshire, Scotland.	Clementina Whyte, wife of a carter.	In a house.	..	Lightning entered by chimney and killed Mrs. Whyte. A child in a cradle close by uninjured.	Times, 26/5/74.
168	About 11 July, 1874.	Victoria Park, London.	Five men.	Walking across the park.	..	One man killed, one had right arm injured. Others received a shock.	Times, 13/7/74.
169	About 11 July, 1874.	Edmonton, near London.	A woman.	Chopping wood in an outhouse.	Chopper	Killed.	Times, 13/7/74.
170	About 11 July, 1874.	Bow, near London.	A man.	With hayfork on shoulder. Had been haymaking.	Hayfork	Killed.	Times, 13/7/74.

125

III. SOME INCIDENTS OF THE ACTION OF LIGHTNING.—(*Continued.*)

126

No.	Time.	Place.	Object.	Particulars.	Lightning rods present.	Metals present.	Results.	Authority.
171	About 11 July, 1874.	Beresford's fields, near Bow, London.	Two lads, Samuel Clarke and Joseph Anderson, and their father.	Haymaking when storm came on, and covered themselves with hay to keep off the rain.	..	..	The two lads killed. The father rendered insensible.	Times, 13/7/74.
172	About 11 July, 1874.	Homerton, near London.	St. Luke's Church.	With a fine stained glass west window.	..	..	Roof rafters set on fire. Stonework of west window shattered.	Times, 13/7/74.
173	18 June, 1880.	Bedminster, near Bristol.	Tree.	Near a tent in which several persons were sheltering.	..	..	Bark stripped from one side. Tree nearly uprooted. Several persons in the tent felt a shock.	W. W. G. 24/6/80.
174	June, 1880.	Clarens, France.	Cherry tree.	Measuring 3 feet 3 inches in circumference.	..	..	Shivered as if exploded by a charge of dynamite.	Times, about 24/6/80.
175	About 11 July, 1874 (6.15 P.M.)	Ayot St. Peter, Welwyn, England.	Parish Church.	..	..	..	Struck, set on fire, and destroyed. (Built 12 years before. Register and papers preserved.)	Times, 14/7/74.
176	About 11 July, 1874.	Cannon Bridge, Tunbridge Wells.	A valuable horse.	Belonging to Mr. Mark, grazing in a field.	..	..	Killed.	Times, 14/7/74.

177	About 11 July, 1874.	Guildford, Surrey.		With chimney stack on north side.	..	..	Chimney stack shattered. Several servants in kitchen struck down, but not much injured.
178	29 May, 1875. (10.30 A.M.)	Valetta, Malta.	Flexford House. (Residence of Mr. D. Onslow, M.P.) R.E. office, and two telegraph stations.	R.E. office used as a military telegraph station and a school for instruction in telegraphy.	*Metals Present.*—(*Continued.*) (3.) Morse recording instruments. (4.) Siemens' protector. (5.) Earth for protector consisting of a copper wire rope of a lightning rod 1 in. diameter, ending in a watertight tank, 4 ft. × 4 ft. × 3 ft. 8 in. full of water.	*At R.E. Office.* (1.) Wires of military telegraph line. (2.) Table in room with ten brass terminals to which line wires were attached. (3.) Iron bar to window of room. (4.) A metallic connection from bar to ground. (5.) A Siemens' plate lightning protector. (6.) Earth for protector formed of gas-pipes in dry ground, connected with gas mains of Valetta. *At two telegraph stations.* (1.) Line wires. (2.) Terminals.	*At R.E. Office.* Discharge was seen to traverse surface of wall of building, between wires and room, like a sheet of flame. A loud report and flash occurred at each terminal, whence discharge passed across room like a ball of fire to window bar, and thence to earth. *At other stations.* The Morse recorders rendered temporarily useless.

III. Some Incidents of the Action of Lightning.—(Continued.)

No.	Time.	Place.	Object.	Particulars.	Lightning rods present.	Metals present.	Results.	Authority
179	1855.	East London, South Africa.	Powder Magazine.	Close to the sea. Sandy soil 12 inches deep overlying old red sandstone. With flag-staff.	Solid iron rod ending in a dry water-tank.	..	Rod struck and torn to pieces. Building much damaged, but no damage to powder.	S.T.E. 12/5/75.
180	13 March, 1844.	Coast of Ireland.	Tower.		..	Iron racer for gun platform. Lead water-pipe leading over a tank, but not touching it.	Truck of flagstaff destroyed, but rest of staff uninjured. Masonry at foot shaken. Masonry between racer and water-pipe pierced. Pipe burst at end over water-tank.	Nels. A.M.
181	7 July, 1872 (night).	Bull Point, near Devonport.	Cookhouse, for use of shell-yard labourers.	Just outside walls of War Department Powder Magazine establishment. Close to river Tamar. On side of a low hill, on summit of which, close to cookhouse, are barracks. Small one-	Particulars—continued. storied building, 25 feet above H.W.M., 33 yards from nearest magazine building with lightning rods, 70 yards from a shell magazine built to hold 15,000 filled shells. Paved floor inside, but not underneath boiler. Narrow pavement outside, round part of building, but not below chimney stack. Built of blue limestone. Brick chimney stack 7 feet high. Earthenware chimney pots 3 ft. high.	Boiler and range. Iron soot door to boiler flue half-way down chimney breast, inside house. Eaves gutters and two down pipes on outside, but not reaching the ground.	Chimney pot of cookhouse knocked down. Chimney stack fractured and soot door knocked out, base of chimney stack not injured. A policeman standing at the door of a small outhouse, 260 yards distant from cookhouse, inside magazine establishment, received a shock during the same storm.	Particulars collected at the place by the Author.

128

No.	Date	Place	Building	Occupants			Effects	Source
182	About 22 Feb., 1880 (evening).	Near Dunmanway, Co. Cork, Ireland.	In a small thatched cottage.	A man named Reily and his sister.	...	...	Killed. Also a dog killed lying in front of the fire. Two children in the room, one on each side of the fireplace, uninjured. Portion of roof destroyed. An aperture pierced in wall. Some fowls outside killed. Wall of kitchen garden broken, and a zigzag furrow ploughed across whole breadth of a 2-acre field beyond.	Times, 24/2/80.
183	About 6 April, 1880.	Chepstow.	Huntfield House.	Residence of Mrs. Stephens.	...	...	Iron gate in front of house. Gilding in rooms (inferred). Chimney stacks battered. A ball of fire passed through dining and two other rooms, blackening everything in its course. Iron gate burst from hinges, and pavement smashed up.	W.M.N. 8/4/80.
184	24 June, 1880.	Barrow Hill Farm, near Ashford, Kent.	House.	Mr. Benton's.	...	...	Chimney struck and shattered. The lightning "passed into a room below, filling it with smoke and sulphur."	K.P. 26/6/80.

III. SOME INCIDENTS OF THE ACTION OF LIGHTNING.—(*Continued.*)

130

No.	Time.	Place.	Object.	Particulars.	Lightning rods present.	Metals present.	Results.	Authority.
185	24 June, 1880.	Eastry, Kent.	Church.	With flagstaff.	..	..	Flagstaff struck. (During divine service.) "The electric fluid passed down close to the clerk."	K.P. 26/6/80.
186	24 June, 1880.	Appledore, Kent.	House.	Mr. Burren Brown's.	..	..	Main chimney struck. Fireplaces disturbed, and various articles in rooms scattered.	K.P. 26/6/80.
187	18 June, 1880 (evening).	Llancant, near Chepstow.	The celebrated "Llancant Elm" tree.	With five sheep under it.	..	..	Elm shattered. Sheep killed.	W.W.G. about 1/7/80.
188	About 18 June, 1880.	Leicester.	St. Matthew's Church.	..	..	Ornamental finial over east window.	Finial struck. Wooden roof ignited. Not much damage.	W.M.N. about 19/6/80.
189	18 June, 1880.	Bristol.	House No. 32, Victoria Street.	Premises of Messrs. Hellyar and Crinks, paperhangers.	..	..	Large hole made in chimney stack. Some coping stones removed. The lightning passed down the chimney into the shop five stories below, and struck a book of patterns in it, and then passed into the street.	W.W.G. 24/6/80.

No.	Date	Place	Object struck	Circumstances			Conductor	Effects	Reference
190	24 June, 1880 (11 A.M.)	East Guildford, near Rye, Kent.	Two men named Wellsted and Chittenden, a boy named Balcombe, and two others.	Had been sheep shearing, and were in a sheep-house having luncheon.	...	...	...	Wellstead, Chittenden, and the boy Balcombe were killed. The others received a shock.	K.P. 26/6/80.
191	23 June, 1880.	Ilkeston.	Post Office telegraph station.	...	...	...	Lightning protector, telegraph apparatus, and wires.	Protector and instruments destroyed.	W.M.N. 25/6/80.
192	13 July, 1880.	Sedgley.	Parish Church.	...	...	...	...	Stonework on south side, and pavement in front of porch, broken in fragments.	W.W.N. 17/7/80.
193	13 July, 1880.	Houghton.	A man.	Sheltering under a tree.	...	...	...	Killed.	W.W.N. 17/7/80.
194	13 July, 1880.	Wolverhampton.	House.	Of the Rev. D. Evans, Baptist Minister.	...	...	...	House struck, and furniture much damaged.	W.W.N 17/7/80.
195	13 July, 1880.	London.	The Ship Restaurant, Charing Cross.	...	...	...	Gutter pipe.	Gutter pipe at top of adjoining house struck; and it fell on a glass roof at the back of the Ship, doing much damage.	W.W.N. 17/7/80.
196	13 July, 1880.	Sheffield.	House.	At outskirts of town, occupied by Mr. Wood, coal merchant.	...	...	Grate.	"Electric fluid passed down the front-room chimney and carried the grate into the middle of the room." Walls of house much cracked.	W.W.N. 17/7/80.

III. Some Incidents of the Action of Lightning.—(*Continued.*)

No.	Time.	Place.	Object.	Particulars.	Lightning rods present.	Metals present.	Results.	Authority.
197	13 July, 1880.	Lower Broughton, Salford.	Sheds.	..	..	..	The lightning struck these sheds, "tearing the roofs off the sheds, and projecting them more than 100 yards."	W.W.N. 17/7/80.
198	17 July, 1880.	Between Pontypool and Newport.	Railway.	..	..	Railway metals.	The passengers in the train were alarmed "as the lightning played near them continually during the journey." A large tree close to the line shattered.	Stand. 19/7/80.
199	24 June, 1880.	Ridden Farm, Lenham, Kent.	A cow.	..	..	..	Killed.	K.P. 26/6/80.
200	24 June, 1880.	The Fostall farmhouse, Tenterden.	House.	Of Mr. James Small, farmer.	..	..	Chimney struck. Lightning entered house. Mr. Small slightly injured.	K.P. 26/6/80.
201	24 June, 1880.	Belgar, Romney Marsh.	House.	Of Mr. G. Gatt.	..	..	"The lightning came through the roof and bedroom, and entered the lower room, two of the inmates of which had a very narrow escape."	K.P. 26/6/80.

Date	Location			Effects		Reference
23 June, 1880.	Netherbridge Farm, Werrington, Launceston.	A horse.	Belonging to Mrs. Lillicrap, standing in farmyard.		Killed.	W.W.N. 25/6/80.
13 July, 1880 (noon).	Ide, near Exeter.	Two cottages.	Small two-storied buildings, built of rubble masonry, plastered and whitewashed, both inside and outside; with one continuous thatched roof over them, and separated by a party-wall. The village consists of one long street, running north and south, on a gentle acclivity; and these cottages are about the middle of the acclivity, and have cottages	*Particulars—continued.* and houses above and below them. The cottages struck are by no means conspicuous or prominent; but the lower or north side of the northern cottage forms an exposed gable end surmounted by a brick chimney stack about 6 ft. high. The upper or south side of the southern cottage is formed by a party-wall separating it from a substantial red-brick house with a slated roof, standing considerably higher. The vacant ground below the north cottage exposed gable end is a vegetable garden. It is elevated some 6 or 7 feet above	*North Cottage.* (1.) Iron kitchen range in north wall. (2.) Iron damper above range. (3.) Iron hinges of bedroom door, rather massive ones, about 9 in. deep, of the HL form. *South Cottage.* Gilt tinsel frames of two small pictures on the back wall of bedroom.	*North Cottage.* Iron damper injured. Small crack in plaster of chimney breast from damper nearly vertically to ceiling. Crack appears again at floor line on north wall of bedroom above, and thence proceeds in a diagonal direction to the lower hinge of the bedroom door, which is a few feet distant, in a partition wall, but close to the north wall. Between the lower and upper hinges, and from the latter to the top of the door (an ordinary ledged batten door) the woodwork of the door was torn away. From the top of the door the crack proceeds straight up the N.E. angle of the room to the ceiling,	Particulars collected by the Author at the place, a notice of the event having appeared in the W.W.N. of 17/7/80.

SOME INCIDENTS OF THE ACTION OF LIGHTNING.—(*Continued.*)

Time.	Place.	Object.	Particulars.	Lightning rods present.	Metals present.	Results.
			the road and above the ground floors of the cottages, which have no basements. In front of the cottages is a narrow strip of irregular pebble paving. Along the backs and the exposed north side there is no paving. The soil in the vicinity is of a fertile nature, and not rocky. The general elevation is low. The village is in a valley; and a			and, just above it, the thatch of the roof, at the gable end, close to the chimney stack, was set on fire, though soon extinguished. The chimney stack was uninjured. The girl says there was a report like a gun going off, but she saw no lightning. Neither she nor the boy were affected in any way. The latter must have been close to the explosion. When their mother came in shortly afterwards she found the house full of smoke and a sulphurous smell. A pane of glass was broken in the front window of the bedroom; but this was probably due to the concussion.

small river, a tributary of the Exe, runs close to it to the southward. There is, however, apparently no water near the two cottages in question. Plenty of trees about. Severe storm. Lasted about an hour. Much thunder and rain.

North cottage has two rooms, a kitchen, on ground floor, and a bedroom above, each being about 10 ft. × 8 ft. × 7 ft. high.

South cottage has two

South Cottage.

(a) The bottle was dashed out of the baby's hand and broken, but the baby was quite uninjured.

(b) The baby's mother, at the other side of the room, was knocked down, but not much hurt; the other two persons were unaffected.

(c) A piece of plaster, about 2 in. square, knocked off the inside of the front wall near the front door.

(d) Another piece, about 6 in. square, knocked off the back wall in one of the back ground-floor closets.

(e) A similar piece knocked off the front wall in the front ground-floor closet.

(f) A patch of plaster, about 2 ft. × 9 in., knocked off north

136

III. SOME INCIDENTS OF THE ACTION OF LIGHTNING.—(*Continued.*)

No.	Time.	Place.	Object.	Particulars.	Lightning rods present.	Metals present.	Results.	Authority.
				similar rooms, and besides, one or two closets on each floor. Each cottage has a small wooden staircase, and a front door opening on to the road. In north cottage, at the time, were a girl and a little boy sitting in the kitchen, the boy being close to the	*Particulars—continued.* kitchen range, which is a fair-sized one with an oven, and has an iron damper in the flue about 2 feet above the chimney bar. The range is in the north or gable end wall. Kitchen paved with lime concrete, but there is brick paving, resting on the ground, under the range. In the kitchen of the south cottage were three persons, and also a baby in a cradle sucking a bottle.*		party-wall in bedroom above. (*g*) Another patch, about 3 in. square, about 3 feet distant from (*f*) knocked off same wall close to ceiling at N.E. angle. (*h*) Some of the gilding on the frames of the two pictures, on the back (or east) wall of this room rubbed off.	

[* There was no apparent connection between stroke at north cottage and that at south; and probably they were two distinct simultaneous explosions. But if it were not for (*h*) it might reasonably be held that there was only one explosion, *viz.* that at north cottage, and that (*a*), (*b*), (*c*), (*d*), (*e*), (*f*), and (*g*) were merely due to the concussion given by that explosion to the two cottages; for there are no cracks on the walls, nor any other marks of the course of the explosion, between any of the eight results mentioned at the south cottage.]

CHAPTER IV.—SOME INSTANCES OF EXISTING LIGHTNING RODS.

No.	Place.	Building.	Prominent features.	Lightning rods.	Remarks.
1	Plymouth.	Guildhall.	Tower 220 feet high, with metal finial.	Copper rope ⅜-inch diameter. Point surmounting finial 10 ft. Glass insulators. Rope kinky and slightly disfiguring. Has 2 rectangular bends in course.	Building erected 1875.
2	Plymouth.	Charles Church.	Granite spire, moderate height, with massive spindle, gilt ball, and vane.	Outside tower, but inside spire, old 1-inch round iron rod. New rod recently erected of ½-inch copper rope, with copper holdfasts, and fixed to base of spindle, which forms point. Old rod disfiguring.	Old church. Spire struck in 1824. See III., 59.
3	Plymouth.	St. Andrew's Church.	Stone tower with pinnacles, each with spindle and gilt vane, 110 ft. high.	Copper rope ⅜-inch diameter alongside, level with, and stayed to, one of pinnacle spindles, with triple point. Disfiguring.	Old church. Earthenware insulators removed this year (1880).
4	Plymouth.	Roman Catholic Cathedral.	(1.) Lofty slender stone spire, with metal finial, cross and weathercock. (2.) Metal cross on east gable summit. (3.) Stone cross on west gable summit.	Copper ropes ⅜-inch diameter. At (2) and (3) projecting 2 ft. Point of (1) not easily visible.	Fine Gothic church, erected 1858.
5	Devonport.	St. Stephen's Church.	Stone spire 160 ft. high, with pointed finial and large gilt weathercock.	Copper rope ½-inch diameter from finial.	Built 1852.
6	Devonport.	St. Mary's Church.	Stone spire 150 feet high, with pointed finial and large gilt vane.	Iron rod ½-inch diameter from finial. Slightly disfiguring.	Built 1850.

IV.—SOME INSTANCES OF EXISTING LIGHTNING RODS.—(*Continued.*)

No.	Place.	Building.	Prominent features.	Lightning rods.	Remarks.
7	Devonport.	Railway station.	(1.) Low turret of passenger station. (2.) Gables of goods shed. (3.) Gables of engine shed.	(1.) Galvanised iron wire rope of three strands, each ¼-inch diameter. (2.) A similar rod at each gable. (3.) A similar rod.	Built 1876.
8	Devonport.	St. James's Church, Keyham.	Stone spire with pointed finial and vane.	Copper rope ½-inch diameter, from finial. Two rectangular bends in course of rope.	Built 1849.
9	Devonport.	St. Paul's Church.	Stone spire with pointed finial and vane.	Copper rod ½-inch diameter from finial. Two rectangular bends in course of rod.	Built 1849.
10	Devonport.	Column.	Tuscan, granite, hollow column, 125 feet high, mounted with a flagstaff 30 feet high, having a spindle and large gilt vane.	Copper tube ¾-inch diameter, running up inside column, and fastened above to foot of one of four ½-inch round iron stays of flagstaff, which reach about half-way up it. Rod has thus no point. Iron hold-fasts to tube inside column much corroded in places. A massive iron balustrade round terrace at top of column, with an iron door leading on to it.	Built 1827 in commemoration of formation of borough. (Very exposed site.)
11	Devonport.	Market buildings.	Stone campanile 124 ft. high, with finial and gilt vane. Lead-covered roof.	Copper tube 1½-inch diameter at base and tapering upwards. Finial makes point.	Built 1825.
12	Devonport.	St John's Church.	Cupola of moderate height, with pointed spindle and gilt vane, resting on stone columns.	Iron rod 1-inch square, running from spindle inside cupola, at foot of which it emerges.	
13	Plymouth.	St. John's Church, Sutton.	Stone spire, with pointed spindle, cross, and weathercock.	Copper rod ½-inch diameter passing from spindle inside spire, and emerging at lattice window of tower.	Modern building.

	Description.	
Christ Church (Unitarian).	Stone spire 80 ft. high, with pointed finial and metal cross.	Iron rod .1-inch diameter at base and tapering upwards. Passing from finial. Rusty and disfiguring.
St. Michael's Church.	Low stone belfry at gable.	Iron rod ¾-inch diameter, point projecting 1 foot, rod ending at rainwater-pipe at eaves. Rusty and disfiguring.
St. Matthew's Church.	West gable.	Iron rod ¾-inch diameter, ending at rainwater-pipe. Point projecting 1 foot.
No. 1, St. Andrew's Place. (Private residence.)	Brick chimney stack at East gable.	Iron rod ¾-inch diameter, dividing above into 3 branches, each with a point projecting 2 feet above chimney tops. Held by wooden holdfasts.
Parish Church.	Stone tower, with turret, finial, and vane.	Copper rod ½-inch square, solid, in 20-feet lengths linked. Point made by finial. Disfiguring.
War Department powder magazines.	Low, vaulted, one-storied buildings, 111 ft. long by 88 ft. wide, with walls of blue limestone, arches of brick, roofs slightly sloping and asphalted, and asphalted side gutters thereon. There are 4 depôt magazines, each holding 10,000 barrels (500 tons), and smaller buildings, together holding powder equivalent to about 7,000 barrels (350 tons); the total amount capable of being stored being thus 47,000 barrels (2,350 tons). The depôt *(Continued.)*	On each of the four depôt magazines the arrangements are as follows:— (1.) At each gable end a vertical copper tube, 2 inches diameter, fixed to walls by copper holdfasts, and with pointed terminal rods projecting 6 feet above the gable summits. (2.) On each long side 4 3-inch copper rainwater-pipes from roof to ground, forming rods, and similarly fixed to walls. (3.) Along roof ridge, joining bases of terminal rods of gable tubes, a flat copper band 4 inches wide, ⅜-inch thick, in centre of which is another pointed terminal rod 8 feet high. *(Continued.)*

IV.—SOME INSTANCES OF EXISTING LIGHTNING RODS.—(*Continued.*)

No.	Place.	Building.	Prominent features.	Lightning rods.	Remarks.
			magazines are 73 yards distant from H.W.M. of the Tamar river, and 25 feet above it. Their front is open to the river, and their rear is enclosed by a high retaining wall. They are built on excavated ground and are isolated from each other by thick earthen mounds. Each depôt magazine has 8 doors and 6 windows, all provided with copper sheathed shutters, not connected to lightning rods. Copper rails (connected to earth) are laid on asphalted roads to convey the powder to and fro, outside the magazines. An asphalted pavement, 6 ft. wide, with brick gutter, surrounds each building. Outside this the yard is covered with grass except where the roads run.	(4.) Crossing the band at right angles at this latter terminal rod is another similar band joining the heads of two of the rain-pipes opposite to each other, on either side. The other 6 rain-pipes, 3 on each side, are not connected with each other nor with the gable rods. (5.) From the feet of the 8 rain-pipes and 2 gable tubes flat copper bands 4-in. by ¼-in. run across asphalte pavement, at edge of which they enter the ground. (6.) Underground, for a total length of 16 feet in each case, they form earth connections, disposed in a forked shape. (7.) The gable tubes also run into small tanks full of water close to their feet.	
20	Plymouth.	Sherwill Congregational Church.	Stone spire, with a massive pointed spindle, cross, and vane, 17 ft. high, removed this year (1880) for a smaller spindle, as the weight (or wind pressure) had displaced the apex stones of spire.	Copper rod about ½-inch diameter, passing down from spindle altogether inside spire and tower, except for a short distance below spindle.	New Gothic church.

No.	Place	Building	Structure	Conductor	Remarks
21	Plymouth.	Millbay Soap Works.	Chimney shaft.	Copper rope ½-inch diameter, with three pronged point, projecting 5 feet.	
22	Devonport.	Keyham Dockyard.	Two chimney shafts.	(1.) Copper rope ¾-inch diameter. Point projecting 3 feet. (2.) Copper tube 3-in. diameter at base, tapering upwards, and dividing into two pointed branches, each projecting 8 feet.	Built 1835. Bakehouse shaft struck in 1841. See III. 96.
23	Stonehouse, Plymouth.	Royal William Victualling yard.	(1.) Clock tower, with copper dome, gilt ball, finial, and vane. (2.) Bakehouse chimney shaft. (3.) Brewhouse chimney shaft.	(1.) Copper tube 3-in. diameter at base, passing inside tower. Finial makes point. (2.) and (3.) Copper tubes 2-in. diameter, points projecting 4 feet.	Modern church.
24	Paddington, London.	St. James's Church.	Stone spire, with no metal above, only a carved stone boss.	Copper rope ¾-inch diameter, projecting 5 feet, and point inclined outwards from boss. Disfiguring.	
25	Torpoint, Cornwall.	Church.	Low metal belfry, with bell, pointed finial and vane.	Copper rod ⅜-in. square, in 6 feet lengths with linked joints, all hanging loose, like a rope, from finial, and broken off about 6 feet above the ground. Very disfiguring.	Old building, apparently un-cared for.
26	Torpoint, Cornwall.	Tor House. (Private residence.)	Chimney stack in centre of roof.	Copper rod ½-inch square, projecting 2 feet.	Old house.
27	Teignmouth, Devonshire.	St. Michael's Church.	Stone tower with corner turret, with spindle and vane.	Iron rod ½-inch diameter, point projecting 2 feet and connected to spindle by an iron stay. Very disfiguring.	Old church.
28	Teignmouth, Devonshire.	Roman Catholic Chapel.	Low stone belfry with stone boss at top, no metal.	Copper rope ½-inch diameter, held by glass insulators. Point projecting 2 feet. Disfiguring.	Built 1854.
29	Dawlish, Devonshire.	Congregational Church.	Stone spire with spindle and vane.	Copper rope ½-inch diameter, with points 6 inches long sticking out from it	Modern church.

IV.—SOME INSTANCES OF EXISTING LIGHTNING RODS.—(Continued.)

No.	Place.	Building.	Prominent features.	Lightning rods.	Remarks.
30	Devonport.	Keyham Gas Works.	Brick chimney shaft.	at intervals of 6 feet along its length up the spire. A brush of 3 points inclined at an angle of 30° to the spindle, and below its summit. Disfiguring.	
31	Newton, Devonshire.	Mackrell's Almshouses.	A range of two-storied buildings. (1.) Central gable over porch, with finial and vane. (2.) and (3.) Chimney shafts at end gables.	Iron rod ½-inch diameter, with porcelain insulators. Prong of 3 points projecting 3 feet. (1.) Copper rope ¾-inch diameter, with earthenware insulators, and a 4 point brush 2 feet below top of finial. (2) and (3). The same, but projecting 2 feet above chimney pots. Rods well fixed and without kinks.	Built 1874.
32	Newton, Devonshire.	Gas Works.	Brick chimney shaft.	Iron rod ⅜-inch diameter, with a 3 point brush projecting 3 feet.	
33	Torquay.	Wesleyan Church, Babbicombe Road.	Lofty stone spire, with ornamental pointed finial.	Copper rope ¾-inch diameter, from finial, fixed by iron holdfasts.	Modern Gothic church.
34	Torquay.	St. Luke's Church.	Low stone spire, with small pointed finial and metal cross.	Copper rope ¾-inch diameter, from finial. Somewhat disfiguring.	Modern Gothic church.
35	Torquay.	Belgrave Church.	Stone tower and pinnacles, each with a metal cross.	Copper rope ¼-inch diameter, at one pinnacle point, projecting 2 feet.	Modern church.
36	St. Marychurch, near Torquay.	Parish Church.	Lofty stone tower with pinnacles, each with a gilt finial and metal cross 10 ft. high, and one with a gilt weathercock.	Copper rope ¾-inch diameter, leading from one of finials.	Handsome modern Gothic church. Tower built in 1871 in memory of Bishop Phillpotts.

No.	Location	Building	Upper structure	Conductor	Remarks
37	Teignmouth, Devonshire.	Pentland Villa. (Private residence.)	Chimney stack.	Galvanised iron rod ½-inch diameter, quite straight and vertical, kept 3 to 6 in. from wall by means of porcelain insulators. Point projecting 2 feet above chimney pot.	Modern house on the side of a hill. Not much exposed.
38	Exeter.	Cathedral.	Two square towers, each with four turrets; each turret surmounted by a massive pointed gilt spindle and vane. Height to tops of turrets 166 ft. Lead-covered roof.	Copper tube 2-in. diameter, down each tower. Each turret spindle has its base connected by a copper band to head of tube at roof of tower. The spindles form the points.	Exposed site. Ground close around walls is not paved.
39	Exeter.	St. Mary Major's Church.	Stone spire with pointed finial and vane.	Copper rope ½-inch diameter, with earthenware insulators, from finial, which forms a delicate brush of 3 points.	Modern Gothic church.
40	Exeter.	Congregational Church.	Stone spire of good height, with pointed finial and vane.	Copper wire band 1½-in. wide and ⅜-in. thick, with glass insulators. Connected to finial above.	Modern Gothic church.
41	Newton, Devonshire.	St. Paul's Church.	Open wooden belfry with a pyramidal slated spire of moderate height, surmounted by a tall metal cross and gilt weathercock.	Copper rope ⅜-inch diameter, with glass insulators, with a 4-point brush 2 feet below top of cross. Disfiguring.	Modern church.
42	Babbicombe, Torquay.	All Saints' Church.	Stone spire of good height, with massive metal cross 12 ft. high, surmounted by a large gilt weathercock.	Copper rope, ⅜-inch diameter, surmounting weathercock by 2 feet with a 3-point brush, and passing down inside spire to lattice window of tower, where it emerges.	Handsome Gothic church, lately built.
43	Torquay.	St. Mary Magdalene's (or Upton) Church.	Lofty stone spire, with metal cross and gilt weathercock.	Copper rope ⅜-inch diameter, with glass insulators, ending in a point alongside and level with bottom of weathercock.	Modern Gothic church.

IV. SOME INSTANCES OF EXISTING LIGHTNING RODS.—(*Continued.*)

No.	Place.	Building.	Prominent features.	Lightning rods.	Remarks.
44	..	H.M.S. *Agincourt.* (Iron-clad battle ship.)	Five masts.	Copper band 3 inches wide, fixed to after side of each top-mast and top-gallant mast. Copper tube 1¾-in. diameter, tarred, passing down one of shrouds from top of each lower mast to side of hull.	Similar system adopted on all H.M. ships.
45	Merrifield, Antony, Cornwall.	Church.	Stone spire with iron cross and gilt weathercock.	Copper wire band 1-inch broad, with glass insulators. Single point inclined at 30° to finial of cross.	Gothic church, lately built.
46	Shaugh, near Dartmoor, Devonshire.	Parish Church.	Granite pinnacled tower, moderate height, no vanes.	Copper rope ½-inch diameter at one pinnacle, single point not projecting above top of pinnacle.	Old church. Struck in 1823. See III. 147.
47	Tavistock, Devonshire.	Congregational Church.	Stone spire surmounted by iron scroll-work, spindle and gilt vane.	Copper wire band 1-inch broad, with glass insulators. Three-pointed prong to rod just below vane, and inclined at an angle. Disfiguring.	Gothic church, built 1872.
48	Tavistock, Devonshire.	New Church.	Stone tower with low stone pyramidal spire and 4 pinnacles. The spire surmounted by a massive iron cross 18 ft. high, and each of the pinnacles by a similar one 12 ft. high.	Copperrope ¾-in. diameter, with earthenware insulators, and surmounted by a prong of 3 points just above top of spire cross.	Handsome new church in Lombardo - Venetian style, built by the Duke of Bedford.

[*Note.*—Unless otherwise stated, the rods pass outside the buildings, have a single point, and are fixed to the walls by holdfasts of the same metal as themselves. The list is compiled from personal inspection, and the dimensions are obtained as accurately as the circumstances would allow.]

Part II.

THE THEORY OF THE ACTION OF LIGHTNING.

CHAPTER V.—ELECTRICAL DEFINITIONS AND DATA.

(A.) ELECTRICAL DEFINITIONS.

(a) *Fundamental Terms.*

(1.) *Electricity* is a temporary state of forced separation of a physical property, normally dormant in all bodies, into two active agencies, each possessed of a power due to a tendency to reunite.[1]

(2.) *Positive electricity,* or *positive charge,* and *negative electricity,* or *negative charge,* are the names given to the two agencies respectively.[2]

(3.) *An electrified body,* or *a charged body,* is one in which either of the electricities is present.

(4.) *Quantity* is the form of agency present with electricity of either nature.[3]

(5.) *Potential* is the form of power present with electricity of either nature.[4]

(6.) *Capacity* is the form of restraint present with electricity of either nature.[5]

[1] I. A *a.*   [2] II. A 19.   [3] I. A 6, 7, 10.
[4] I. A 6, 7, 10.   [5] I. A 6, 7, 10.

N.B.—These footnotes are references to other Chapters, Sections, and Paragraphs.

V. A 7—17.

(7.) *Attraction* is the property which electricity of one nature has for that of another.

(8.) *Repulsion* is the property which electricities of the same nature have for each other.[1]

(9.) *A collector* is a body quâ its property of permitting electricity on it to distribute itself, and hence to accumulate.[2]

(10.) *An insulator* is a body quâ its property of preventing electricity on it from distributing itself, and hence from accumulating.[3]

(11.) *A conductor* is a body quâ its property of permitting electricity to be transmitted through it.[4]

(12.) *Discharge* is the act of reunion of two electricities of different natures, involving the execution of work, and the restoration of the bodies on which the charges were collected to their normal passive state.

(13.) *An explosion* is an instantaneous discharge through an insulator.[5]

(14.) *A return stroke* is an instantaneous discharge through a collector, occasioned by a reunion of the two electricities originally separated.[6]

(15.) *A leak* is a continuous discharge through an insulator.[7]

(16.) *Electro-motive force* is the force manifested in the obliteration of difference between the potentials, or the states of potential, of two electrified bodies when joined by a conductor.[8]

(17.) *A current* is a transmission of electricity by electro-motive force through a conductor; and it comprises all discharges through a conductor, whether instantaneous or continuous.[9]

[1] I. A 2.
[2] I. A 15.
[3] I. A 14—16.
[4] I. A 15.
[5] I. A 27—31.
[6] I. D 51.
[7] I. A 16.
[8] I. A 8, 13.
[9] I. A 8, 13.

V. A b c 18—30.

(b) *The Influence of Bodies.*

(18.) *Influence* is the inherent property, manifested by various actions, that bodies possess for affecting electricity.

(19.) *Collection* is the action of distributing electricity and of imparting by it electricity of the same nature.[1]

(20.) *Induction* is the action of imparting one kind of electricity by means of the other.[2]

(21.) *Insulation* is the action of isolating or enveloping electricity, and of limiting its distribution.[3]

(22.) *Facilitation* is the action of hastening explosion, and of transmitting its operation without obstruction.[4]

(23.) *Restraint* is the action of delaying explosion and of obstructing its operation.[5]

(24.) *Conduction* is the action of transmitting current.[6]

(25.) *Resistance* is the action of retarding current.[7]

(c) *The Nature of Condensers.*

(26.) *A condenser* is such a juxtaposition of two collectors, one of which is electrified from some extraneous source, that by means of mutual induction through the separating insulator the potential of the electricity is appreciably raised.[8]

(27.) *A collecting plate of a condenser* is the surface, nearest to the insulator, of the collector which receives its original charge from some extraneous source.

(28.) *A condensing plate of a condenser* is the surface, nearest to the insulator, of the collector which receives its original charge by induction from the other collector.

(29.) *A dielectric of a condenser* is the insulator separating the collecting from the condensing plate.[9]

(30.) *An explosion of a condenser* is an instantaneous dis-

---

[1] I. A 15.   [2] I. A 17—21.   [3] I. A 14—16.
[4] I. D 2, 6.   [5] I. A 27. I. D 17.   [6] I. A 15.
[7] I. A 8.   [8] I. A 21—26.   [9] I. A 24.

148   LIGHTNING.

V. A 31, 32; B *a* 1.

charge, through the dielectric, of the electricities accumulated on the plates.[1]

(31.) *A return stroke in a condenser* is an instantaneous discharge of the electricity on one of its plates, through the plate itself, owing to the electricity on the other plate having become discharged by some extraneous means.[2]

(32.) *A leak in a condenser* is a continuous gradual discharge, through the dielectric, of the electricities accumulated on the plates.[3]

(B.) Electrical Data.

(a) *Electrical Formulæ.*

(1.) The following formulæ express mathematically the chief laws on which electrical science is based.[4]

I.—Fundamental.
Mass $= m$.
Distance $= l$.
Time $= t$.

II.—Mechanical.

$$Velocity = v = \frac{\text{distance}}{\text{time}} = \frac{l}{t}.$$

$$Acceleration = a = \frac{\text{velocity}}{\text{time}} = \frac{l}{t^2}.$$

$$Force = f = \text{acceleration} \times \text{mass} = am = \frac{ml}{t^2}.$$

$$Work = w = \text{force} \times \text{distance} = fl = \frac{ml^2}{t^2}.$$

III.—Electrical.

$$Quantity = q = \sqrt{\text{force}} \times \text{distance} = f^{\frac{1}{2}}l = \frac{m^{\frac{1}{2}} l^{\frac{3}{2}}}{t}.$$

[1] I. A 27, 29, 30.
[2] I. D 51—58.
[3] I. A 16, C 88. II. D 11.
[4] I. A 6—11.

$$Potential = p = \frac{\text{work}}{\text{quantity}} = \frac{w}{q} = \frac{m\,l^2}{t^2} \times \frac{t}{m^{\frac{1}{2}}l^{\frac{3}{2}}} = \frac{m^{\frac{1}{2}}l^{\frac{1}{2}}}{t}.$$

[Hence potential $= \sqrt{\text{force}}$.]

$$Capacity = s = \frac{\text{quantity}}{\text{potential}} = \frac{q}{p} = \frac{m^{\frac{1}{2}}l^{\frac{3}{2}}}{t} \times \frac{t}{m^{\frac{1}{2}}l^{\frac{1}{2}}} = l.$$

[Hence capacity = distance.]

Electro-motive force $= e =$ difference of potential $= p$.

$$Strength\ of\ current = c = \frac{\text{quantity}}{\text{time}} = \frac{q}{t} = \frac{m^{\frac{1}{2}}l^{\frac{3}{2}}}{t^2}.$$

$$Resistance = \frac{\text{electro-motive force}}{\text{current}} = \frac{e}{c} = \frac{m^{\frac{1}{2}}l^{\frac{1}{2}}}{t} \times \frac{t^2}{m^{\frac{1}{2}}l^{\frac{3}{2}}} = \frac{t}{l}.$$

(b) The Three Elements of Electricity.

(2.) Quantity, potential, and capacity are the three essential elements of electricity.[1]

(3.) The basis of electrical law is the expression denoting quantity, which is derived experimentally from the amount of force of repulsion or attraction developed between two small electrified bodies placed at a certain distance apart.[2]

(4.) Potential is the impelling or moving quality of electricity. It is the measure of the capability that electricity has of doing work, in proportion to its quantity.[3]

(5.) Capacity is the restraining or limiting quality of electricity. Its action is directly antagonistic to that of potential. It is the element that allows of the electricity being tangible.[4]

[1] I. A 7.   [2] I. A 6.   [3] I. A 10.   [4] I. A 10.

V. B *c d* 6—11.

(*c*) *Collectors and Insulators.*

(6.) Every electrical system consists of two collectors separated by an insulator.[1]

(6*a*.) Every electrified body consists of a collector enveloped by an insulator.[2]

(7.) The capacity of an insulator consists of three factors, and varies directly as each of them, viz.:—

- (*a*) *Surface*, or the extent of area of the insulator, *quâ* the collector.[3]
- (β) *Thickness*, or the distance of the surface of the collector from that of the nearest collector from which it is separated by the insulator.[4]
- (γ) *Restraint*, or " specific inductive capacity ; " *i.e.* the specific influence due to the material of the insulator.[5]

(8.) A body which has one of the three characteristics of being a quick collector, a great facilitator, or a good conductor, necessarily possesses the other two characteristics.

(9.) A body which has one of the three characteristics of being an insulator, a restrainer, or a non-conductor, necessarily possesses the other two characteristics.

(*d*) *Electrical Explosions.*

(10.) The potential and capacity present with any electrified body always balance each other, and the loss of balance constitutes discharge.[6]

(11.) An explosion of a condenser occurs when the potentials of the charges separated on the two plates have accumulated to such a degree that the capacity of the dielectric is unable any longer to restrain their junction, *i.e.* to balance the combined effect of the two potentials.[7]

[1] I. A 6.   V. A 4, 6 ; B 3.   [2] I. A 20.   V. A 6.
[3] I. A 25, 26.   V. A 23.   [4] V. A 6 ; B 5.
[5] I. A 6.   V. A 6 ; B 5.   [6] I. A 7.
[7] I. A 29, 30 ; D 11, 16, 17.   II. C 27 ; V. A 13.

(12.) From the formula capacity $=\frac{\text{quantity}}{\text{potential}}$,[1] it follows that potential $=\frac{\text{quantity}}{\text{capacity}}$; hence potential is liable to be affected by alterations to quantity and capacity; but since capacity varies as surface, thickness, and restraint, and these conditions cannot affect, or be affected by, the quantity of electricity present, it is evident that, in any given electrical system, alterations to quantity, surface, thickness, or restraint, will always alter potential.

(e) Electrical Return Strokes.

(13.) If one of the plates of a condenser becomes discharged by means of explosion with a plate of an adjacent condenser, so much of the charges on each of the remaining plates of the two condensers as was due to the induction of the discharged plates necessarily reunites through itself with the complementary electricity from which it was separated by the action of induction.[2]

(14.) This action of return, or recombination, is of a nature not unlike that of a current, and constitutes a return stroke or induced discharge.

(f) Electrical Leaks.

(15.) In practice, perfect insulators do not exist; and there are two kinds of leaks to which all insulators, and hence all dielectrics, are subject.[3]

(16.) In the first place, they all leak, in inverse proportion to their specific restraint, through the electric pores of their material.[4] Hence, in every condenser, the charges on the two plates are continually, slowly, and silently rejoining each other, but frequently in so comparatively small a degree that the general conditions of the condenser

---

[1] I. A 10. V. B 1. [2] I. A 51—58, 66—71. V. A 14.
[3] I. A 16. V. A 15. [4] V. B 7.

V. B g 17—21.

are perhaps not, till after a considerable time, materially affected.

(17.) Secondly, they leak through the indentations on their surfaces caused by projecting angularities on the surfaces of the collectors they envelope, and in direct proportion to the number and acuteness of these angularities.

(18.) Thus, wherever a point, angularity, or indeed any salient deviation from a uniform surface, occurs on the face of a collector, there, in proportion, capacity ceases, potential becomes infinite, and discharge occurs,[1] although the rate at which electricity is thus lost may be extremely slow.[2]

(19.) If the collector be so shaped as in itself to constitute an acute angularity, or point, and if it moreover consist of highly collective material, as metal, the entire charge on it quickly disappears.

(20.) The inherent power of a point on a collector to eject charge therefrom is quite independent of the proximity of any juxtaposed collector.[3]

(g) *Illustrations of Electrical Action.*

(21.) It may serve to strengthen the ideas if we illustrate the leading elements of electro-static action by means of hydraulic engineering similes.[4] The following table shows this comparison:—

Electric Element.	*Hydraulic Simile.*
Condenser	Reservoir formed by a dam constructed across a gorge.
Collecting and condensing plates.	Bottom and sides of the reservoir.
Dielectric	Dam retaining the water in the reservoir.

[1] V. B 10.   [2] II. C 11, 17, 26—28, 37—39, 51, 53—55.
[3] II. C 39, 51, 53—55.   V. B 15.   [4] I. A 12.   II. C 51.

ELECTRICAL DEFINITIONS AND DATA.

Electric Element.	*Hydraulic Simile.*
Charge on the plates	Body of water in the reservoir.
Potential of the charge . . .	Head of water in the reservoir.
Capacity of the dielectric . .	Dimensions and material of the dam.
Discharge of the plates . . .	Escape of the water by any means.
Explosion of the condenser . .	Bursting of the dam through the pressure of the water.
Porous leak in the condenser .	Porosity of the material of the dam.
Angular leak in the condenser.	Sluice, or pipe, through the dam.

VI. A a.

CHAPTER VI.—THE CONSTITUTION OF THE TERRESTRIAL CONDENSER.

(A.) THE FUNCTION OF THE EARTH IN THE TERRESTRIAL CONDENSER.

(a) *The Relation between the Earth and the Clouds.*

DURING, and immediately previous to, a thunderstorm, the earth, atmosphere, and clouds, constitute a great natural electric condenser, of which the earth and clouds are the two plates, and the intervening atmosphere is the dielectric; and the explosion of this condenser is manifested under the form of shafts of lightning, accurately termed thunderbolts.[1]

It is of importance that we should have a clear perception as to which of these two plates it is that collects the electricity, and which condenses it.

According to the theory generally accepted, the clouds collect or originate the electricity of thunderstorms, whilst the earth acts as the condensing plate.[2]

It is assumed that the clouds receive their charge by collecting from the upper regions of the atmosphere the electricity perpetually being evolved from the surface of the earth through decomposition, evaporation, and other chemical actions,[3] and that the earth receives its charge only by induction from the clouds.

The theory that the earth forms the condensing plate of the terrestrial condenser is probably based on the fact of

[1] I. D 1, 2, 11, 15, 16. II. E 1. V. A 26—30; B 11, 21.
[2] I. D 11; C 49. II. C 43.   [3] I. C 10—14, 34.

VI. A a.

the necessary arrangement of artificial condensers, wherein the condensing plate is connected to the earth, by which means the electricity on this plate, complementary of the charge induced by the collecting plate, can be driven into the earth, thus affording room for the induced charge to attain its maximum potential.

But the principle of the condenser is clearly that of induction, and does not depend on the fact of the condensing or induced plate being connected with the earth;[1] in other words, it does not depend on the earth itself forming the condensing plate.

Doubtless an artificial condenser of any practical value could hardly be formed unless the condensing plate were thus formed by the earth; but in the case of the great natural condenser of which the earth directly constitutes one of the plates, and the clouds the other, the circumstances are not the same.

All telegraphic and electric experience tells us that the earth itself is a great reservoir of electricity,[2] and such being the case, it would seem only natural that the electricity in this reservoir should occasionally be collected at places on its surface; but if this should happen, it is necessary by electric law that charge of an opposite nature should be induced on the nearest separate collector.[3]

Now the nearest separate collectors to the earth's surface during the existence of ordinary thunderstorm conditions are certainly the clouds. Hence, supposing that, owing to the action of the electricity within the earth, charge should accumulate at places on the earth's surface, charge of an opposite nature must always be induced on the surfaces of the clouds nearest to these places; and thus with the earth's surface as the collecting plate, the nearest clouds as the condensing plate, and the insulative regions of air between them as the dielectric, we should obtain the terrestrial condenser.

---

[1] I. A 21.   [2] I. C 52—55.   [3] I. A 21. V. B 7.

VI. A b.

But there is good ground, based on the phenomena of auroræ,[1] of heat (or sheet) lightning,[2] of lightning attending volcanic eruptions,[3] and of thunderbolts occurring in clear skies,[4] for believing that the upper regions of rarefied air do actually (owing probably to induction by the earth) form to some extent a reservoir of electricity independent of the clouds, and any conduction of electricity upward from the earth's surface, due to the process of chemical evolution, would doubtless tend to promote the formation of such a reservoir.[5]

All the functions that the earth fulfils towards artificial condensing plates would thus be fulfilled by this atmospheric reservoir towards the clouds; and these would be enabled to act efficiently as condensing plates to the great collecting plate of the earth's surface.[6]

The fact that the greater proportion of lightning occurs between cloud and cloud rather than between cloud and earth,[7] will not militate against the idea of the clouds receiving their charges by induction from the earth, if it be borne in mind that two of the great observed elements of the clouds are their detachment from each other in separate masses, and their mobility.[8]

Their separateness permits them to hold distinct insulated charges, and to form fresh condensers with each other;[9] whilst their mobility occasions more chances of their being driven within explosive range of each other than of the earth, which relatively is fixed.

(b) *The Earth's Electricity.*

We appear to have *primâ facie* grounds for believing that the earth's surface is really the collecting plate of the terrestrial condenser, from the fact, already mentioned, of

---

[1] I. C 77, 78.   [2] I. D 4, 20.   [3] I. C 114, 115, 118—120.
[4] III. 1, 2, 50. I. C 4.   [5] I. B 4; C 35.
[6] I. C 34.   [7] I. D 16. VI. B a.
[8] I. C 31, 33.   [9] I. C 33.

VI. A b.

electricity being contained within the earth; but the questions now arise, what is the original source of the earth's electricity, and how does its surface collect it?

In our present state of knowledge it seems to be impossible to get beyond conjecture in replying to such questions.[1]

Supposing, however, that we take up the opposite view that the clouds form the collecting plate, the task of attempting to prove how they originate and collect their electricity would appear to be even more hopeless; for, although we reasonably infer that the clouds are collectors of electricity, we do not know the fact for certain; and *d fortiori* we must be ignorant of how the electricity got there.

But we do know with certainty several important facts regarding the earth's electrical constitution; one is (as already mentioned) that it is a great holder of electricity;[2] a second, that it is a great collector and conductor of electricity;[3] a third, that portions of its surface are constantly found to be electrified;[4] a fourth, that it is a great magnet;[5] and a fifth, that terrestrial disturbances, such as waterspouts, earthquakes, and volcanic eruptions, are connected with the actions of electricity or magnetism.[6]

We have thus, in our investigations as to the probability of the earth being the originator of thunderstorms and lightning, several scientific data to work upon, which are quite wanting in regard to the clouds; and we propose now to make a few remarks on the possible nature of the earth's electrical actions, with the view of strengthening the likelihood of the theory that its surface is the collecting plate of thunderstorm condensers.

As to how the earth obtains its electricity or how it became a magnet, we are practically in total ignorance.[7]

---

[1] I. C 2, 71—73. [2] I. C 52, 53. [3] I. C 50, 51.
[4] I. C 6, 9.   [5] I. C 63.   [6] I. C i, k, l.
[7] I. C 71—73b.

VI. A b.

The fact, however, that it is simultaneously both a holder of electricity and a magnet is well worthy of attention; and so also is the fact that phenomena undoubtedly electrical, *i.e.* earth currents and auroræ, are invariably accompanied by magnetic disturbances.[1]

It is well known that what is generally called magnetism is so closely allied with electricity that the one produces the other, that in some respects the action of magnetism is the same as that of an electro-static charge, or of electricity at rest, whilst in others its action is the same as that of electricity in motion.

Owing to this last fact, magnetism has already been conceived to be a series of electric currents, and the earth's magnetism has been considered to be due to the never-ceasing motion of these currents around it.[2]

But may we not take a simpler view of the matter and conceive the subtle force usually called magnetism to be nothing but electricity, *i.e.* electricity bound or manifested in a peculiar manner, and magnetism itself as only a property or influence analogous to conductivity or inductive capacity, appertaining to certain bodies, and permitting this particular manifestation?

All the special characteristics of magnets could apparently be comprised by such a theory, for electric currents cannot exist without electricity; in other words, electricity in motion does not cease to be electricity.

The actions of magnets would then be considered as electrical actions, due to their occurring on a magnetic body, on the same principle as other actions of electricity are due to the fact of their occurring on insulating or conducting bodies.

On this principle, then, the earth is a magnetic body like steel and iron, and what is known as its magnetism becomes an additional proof of the presence and activity of its electricity, and strengthens the probability that the globe is

---

[1] I. C 56, 57, 60a, 60b, 64, 73, 75, 81, 82.   [2] I. C 68—71.

VI. A *b.*

itself the originator of thunderstorms and of all other electrical phenomena known to occur in connection with it.[1]

That the separated agencies composing this electricity should be in constant motion in the magnetic field or orb, from the equator towards the poles, is what is to be expected from the analogy of the action of the lines of force of magnets; hence, in the field of the earth-magnet, which we know also to be a conductor, we have manifestations of motion (rendered irregular by induction and geological causes) in the shape of earth currents.[2]

The polar accumulations of electricity thus formed would influence by their powers of attraction and repulsion other electricities similarly accumulated, within their range or field;[3] and this would explain the attraction of the earth's poles on those of other magnets.

To the same fact of dense accumulation of electricity at, or near, the magnetic and terrestrial poles, coupled with the general absence of rain-clouds thereat, would be attributed the manifestations of silent continuous discharges, or leaks of electricity, in the polar regions, seen under the form of auroræ.[4]

And it is conceivable that the electricities, in their motions through or over the earth towards the poles, are occasionally forced by obstructions, due to geological conditions, to accumulate for a time at certain places on the surface; and when this should occur in regions where clouds and rainfall were frequently present, the necessary conditions for the development of thunderstorms would apparently be obtained.[5]

Lastly, if the occasional irregular accumulations should occur in certain portions of the earth's crust (generally not very far distant from the sea), adjacent to, but insulated from, each other, and below, though not far removed from, the surface,—and especially in regions where clouds and

---

[1] I. C 110; G 22.   [2] I. C *e.*   [3] V. A *a.*
[4] I. C *g.*   I. G 25.   [5] I. C 15—17, 34.   I. D 11, 15, 18.

VI. B a.

rainfall were habitually absent, as in Chili and Lower Peru,[1]—there would appear to be possible causes for the occurrence of earth explosions, manifested by earthquakes,[2] and their frequent volcanic accompaniments;[3] it being borne in mind that these convulsions have frequently been found to coincide in time with magnetic fluctuations, and that eruptions are constantly associated with lightning.

Taking, then, into account all the circumstances of the two hypotheses concerning the function of the earth in the terrestrial condenser, the one assuming that the clouds originate thunderstorm electricity, and that the earth condenses it, and the other, that the earth originates this electricity, and that the clouds condense it—the latter appears the more reasonable of the two.[4]

We shall, it is hoped, see later on in this treatise, whilst discussing the subject of *leaks*, that faith in the idea that the earth's surface is the collecting plate of the terrestrial condenser considerably facilitates the task of devising practical measures for attempting to prevent the occurrence of thunderbolts.[5]

(B.) The Theory of Descending Lightning.

One great reason for the assumption that the clouds are the originators of the electricity of thunderstorms has probably been another assumption, and one apparently of nearly universal acceptance, viz. that lightning strikes the earth from above.[6]

We propose to examine this theory, first, from the point of view of fact; and secondly, from that of electrical law.

(*a*) *Facts regarding Descending Lightning.*

Lightning appears to be an act, or manifestation, too vivid and instantaneous to permit of the eye doing more

---

[1] I. G 23, 24, 28. [2] I. C *k*; G 22. [3] I. C *l*.
[4] I. C 4. [5] VII. G *c*. [6] I. D 1, 12, 14, 16. II. E 35.

VI. B a.

than merely noting its existence. It seems quite impossible, from sight, to say positively whether lightning ascends or descends, or has any motion at all. It would probably be in strict accordance with reality to say that a luminous rift or crack suddenly appears and disappears in the gloomy atmosphere before the brain has time to inform the mind of anything more than these bare facts.[1]

As regards the traces lightning leaves behind it, the writer has as yet been unable to meet with any single record tending to prove that a stroke of lightning had acted as a downward force.

The circumstances, however, of the strokes recorded in the following eighteen incidents in Chapter III., tend to show, more or less strongly, that in each case the action of the explosion was upwards, viz. :—

Nos. 5,	25,	103,
11,	26,	112,
12,	29,	124,
15,	40,	137,
16,	51,	173,
24,	62,	197.

In No. 5 the evidence of the column of water shot upwards into the air is remarkable.

In Nos. 11, 12, and 62, stones were thrown to distances of 50, 55, 60, and 400 yards, the stones thrown 50 and 60 yards weighing respectively 70 and 336 lbs.; in No. 112 a piece of wood 10 feet long was hurled to a distance of 50 yards; in No. 197 the roof of a shed was projected more than 100 yards; and in No. 16 a timber spire of a church was thrown to some distance.

Can we reconcile these circumstances with the idea of a force acting from above? From the weights and sizes of the bodies, and the distances to which they were propelled, it seems clear that the force must have been upwards.

[1] I. D 8, 28, 29.

VI. B a.

The usual explanation of any great uplifting force manifested by a lightning explosion appears to be that the place struck must have contained more or less moisture, the sudden conversion of which into steam by the intense heat of the stroke has caused the effects in question.[1]

But this is entirely a supposition; and certainly, in the instances above quoted, there is no record of the presence of moisture.

It is of course possible that the sap in trees may become thus transformed into vapour; but it is difficult to understand how the interior of a flagstaff or of a stone wall could contain moisture to any extent.

Such a theory would moreover appear to involve the occasional rending of the ground in cases where lightning is supposed to "make good earth," *i.e.* to strike moist ground; but such results never appear to occur in these instances.

In No. 15, the fact of the horses' hair being burnt on the legs and under the belly is noteworthy.

Lord Mahon, a distinguished physicist,[2] in a report at the time on this case to the Royal Society, considered it to be a return stroke, but Arago says that it showed "undeniably the principal effects of an ordinary stroke of lightning."[3]

In Nos. 16, 24, 51, and 173, there is clear evidence of an uprooting force; and in No. 51 we are informed in Sir William Snow Harris's own words, that the clock-room floor was left "as if blown up by gunpowder."

Nos. 25, 26, 29, and 124, appear to be almost as direct evidences of the upward exertion of the force as it is possible to obtain.

Nos. 40, 103, and 137, also furnish, in their cases respectively, fair grounds for belief in the existence of a similar force.

It may perhaps be doubted whether these few instances

---

[1] I. D 44.   [2] II. A 28.   [3] VII. F c.

VI. B a.

we have quoted are by themselves sufficient vouchers for the existence of a force in lightning strokes invariably acting upwards; but it is extremely probable that the popular belief to the contrary must have tended to prevent any detailed examination of the traces of thunderbolts by the light of this theory, the result of which belief has been that, as a rule, the most meagre details are given of the action of the explosion, except where the effects were shown on the building or person, and the particular points of ground where the lightning was supposed to have dispersed have been but little noticed.

It may be mentioned that in only five of the cases given, viz. Nos. 24, 25, 26, 29, and 124, have the facts been published expressly to denote the existence of ascending lightning.

The 203 incidents mentioned in Chapter III. have been collected without reference to any particular theory, and merely in the hope of throwing light on the action of lightning; and it is now suggested that the reader should experimentally peruse any one of them under the idea that the lightning ascended, and sprang from the place (if such is mentioned) where it was presumed to have entered the ground, bearing in mind that the sequence of events recorded in the column of results would usually be inverted (the description employed by the authority for the incident having generally been adopted), and would always commence at the ground, and thence proceed to the topmost feature of the construction or object under consideration.

We venture to think that the result will be that nothing will be found in the facts related that is in disaccordance with the theory of ascending lightning.

We will now consider the question from a mechanical point of view, in connection with the striking of the points of lightning rods by lightning.

The mechanical force present in lightning is supposed to descend from above, to alight on the top of the unstayed projecting slender metal terminal rod (say $\frac{1}{4}$ inch to 1 inch

VI. B *a*.

in diameter, and 3 to 30 feet[1] high) of a presumably efficient lightning conductor "making good earth," then to proceed down it to earth, all without effecting any mechanical injury; and, in most cases, only, at the utmost, to manifest its presence by fusing a small portion off the point of the terminal.

But if this enormous force really does strike in this way, surely it ought either to break off the terminal at its junction with the part of the rod that is fixed to the wall, or to knock the whole conductor down and to tear the holdfasts out of the wall.

It may be said that the rod being of metal disperses the discharge by affording it an easy electrical passage to the ground, and thus keeps itself uninjured.

Allowing, for argument's sake, that this is the case, we may still ask what becomes of the mechanical force with which the lightning arrives at the rod?[2]

There is apparently no electrical law by which conductors of electricity may receive a blow directed at them from the outside without feeling its effects, by reason of this blow having been itself actuated by electricity.

Now take the case of the same efficiently erected rod under the idea of its receiving from the ground the electricity that actuates the blow. The circumstances are changed. The rod does not receive the blow, but transmits it; the reaction is on the ground; and there is much less reason why the slender unsupported terminal should be broken, or why it should show any manifestation of the thunderbolt's passage beyond traces of the intense heat it must inevitably be subjected to when the discharge is on the point of leaving it and becoming lightning.

Incidents Nos. 6, 8, 22, 56, 64, 66, 78, 79, 89, 92, and 156, are instances of lightning rods being struck and receiving no material injury.

---

[1] II. C 2, 3, 16, 23, 24, 48—50. IV. 1, 15, 16, 19, 21—24, 26—28, 30, 32, 35, 37, 41, 42.   [2] VII. C 14. VIII. C 1.

VI. B b.

(b) *Descending Lightning from the Aspect of Electrical Law.*

As to the scientific side of the question, doubtless one of the reasons for maintaining th lightning usually descends to the earth, has arisen from the fact that the electricity on the earth's surface during fair weather has been found to be usually of a negative kind;[1] hence it has been inferred that the same holds good during thunder weather, and that the electricity of the clouds is then of a positive nature, and that on analogy from the hypothesis that a current traverses a conductor in the direction from positive to negative charge, the direction of the lightning discharge would generally be from the clouds to the earth.[2]

Premising that we appear to have at present no definite knowledge of the nature of the electricity resident during thunderstorms on the earth or in the clouds (except that, from the fact of explosions occurring between clouds, it is evident that they may possess either kind of charge), the above reason would seem principally to fail owing to the obvious fact that lightning strokes, and electric sparks generally, are evidences of explosion and not of current.[3]

There is apparently no ground for assuming that the action of an explosion follows the same law as that of a current.[4]

On the contrary, inasmuch as the scene of the one is formed by bodies which have an exactly contrary influence to those which form the scene of the other, it is only reasonable to presume that the action of the one is also very different from that of the other.[5]

This hypothesis is strengthened by the fact that, in all lightning explosions, the course of the discharge through any metals that happen to be in the dielectric does not follow the longest dimension of these metals, as it would if they were conducting it like a current, but simply utilises as

[1] I. C 5.
[2] I. D 12. II. C 43.
[3] I. A 27; D 11, 16. V. A 13, 17, 30.
[4] I. A 27.
[5] V. A 13, 17. VI. D a.

VI. B b.

stepping-stones such portions of them as happen to lie in its path of least restraint to the condensing cloud's electricity, and leaps over any intervals occurring between them; and their influence on the discharge appears to have exactly the same result as if they attracted it.[1]

In the case of an electric spark leaping across a small gap in a conductor, through which a strong current is passing, the electricity is certainly brought to the edges of the gap on each side, by means of the current; but there the work of the current ends, and what makes the spark pass across is the formation of a condenser, and the explosion of it resulting from the potential accumulated on either side of the gap.[2]

In the case of the lightning electric spark we have two electricities, positive and negative, lying respectively on either side of the great gap formed by the atmosphere. The combined force of the two potentials has accumulated to that stage where the capacity of the gap is no longer able to restrain their fierce embrace; but there is no reason for supposing that positive has at this time more attraction for negative than negative has for positive.[3]

The conclusion, then, seems irresistible that, if there is any element of time in the case at all, the lightning spark leaves the two plates, the earth and the clouds, simultaneously, and coalesces half-way between; and this would result in an invariable upward direction of the stroke immediately above the surface of the ground.[4]

Probably mythological traditions have had a considerable share in the formation of the belief that thunderbolts strike the earth from the skies.

It is submitted, then, that the idea of the clouds being the originators of thunderstorm electricity is devoid of foundation so far as it rests on the theory that lightning descends.

VI. C.

(C.) THE OUTLINE OF THE TERRESTRIAL PLATE.

Independently of the question whether the collecting plate of the terrestrial condenser is formed by the earth or the clouds, it is of importance that the exact outline of the terrestrial plate should be determined; for it is impossible to deal satisfactorily with the subject of protecting life and property from the ravages of thunderbolts until a definite opinion has been formed on this point.

The theory on it of most recent acceptance appears to be this, viz. the outline of the terrestrial plate is the uppermost surface of the most collective stratum of the earth's crust.[1]

By this theory it follows that the rocks and less collective portions of the earth's surface overlying this more moist and collective stratum, form, equally with the air, parts of the terrestrial dielectric.[2]

Since, however, those portions of the earth's crust which are not actually rock, and are consequently more liable to receive moisture, *e.g.* earth, clay, loam, and sand, are merely disintegrated rock, it is evident that all portions of the earth's crust are allied in physical composition more nearly to each other than to air, and that, therefore, *cæteris paribus*, the actual rocks are the less likely to become electrically separated from their moister earthy surroundings and to join the air in forming the dielectric of the condenser; and especially so when it is borne in mind that stones and rocks can have but little inductive capacity.[3]

Assuming, however, that the rocks on the earth's surface may actually have inductive capacity sufficient to form part of the dielectric of a condenser, it is certain that their restraining power would be considerably less than that of the air, and, by so much as would be due to that fact, would facilitate discharge from the collecting plate below them.[4]

[1] I. C 47, 48. II. D 44. [2] V. A 26—29.
[3] VI. D *a*.   [4] VI. D *a*.

VI. C.

Rocky masses overlying moist strata would thus rather encourage discharge; and in the open country these strata thus overlaid ought not infrequently to be the scenes of thunderbolts; and as we know from experience that the action of explosion passing through substances of the same nature as rocks, *i.e.* stone spires and walls, is of a mechanically expansive, rending nature, it would follow that sometimes, during thunderstorms, rocky strata would be rent in pieces, mountains would be uprooted, houses and even towns would be thrown down; and the effects of thunderbolts would, in fact, occasionally resemble those of earthquakes.

But is this the case? Do not our records tell us in the first place that, as a rule, according to the generally adopted form of expression, "lightning seeks good earth?"[1] And what is this but telling us, in other words, that discharges do not generally occur at rocky surfaces at all, wherever moist surfaces are adjacent, and other things are equal?[2]

And, in the second place, even at the rocky summits of high mountains, where the facilitating element of great elevation so overpowers the restraining one due to the rocky surface that thunderbolts are not uncommon, do we ever hear of crags and peaks being torn asunder by lightning?[3]

But the stone walls of buildings being of the same nature as rocks would facilitate explosion equally, and the massive stone walls and arches of powder magazines ought to be sources of danger to the powder, whilst the brick and stone walls of dwelling-houses would tend to bring destruction rather than protection to their inmates.

The presence of rain during thunderstorms, so far as it would affect the question, would appear to be entirely against the theory of the rocks on the surface forming part of the terrestrial dielectric, since, during rainfall, it is

---

[1] I. C 45.   [2] VII. A *d*.   [3] I. G 33, 41. VII. A *g*.

VI. C.

obviously the surface of the earth that first, and in the greatest amount, receives moisture, and some time must necessarily elapse before any lower strata receive the access of collectivity due to this cause.[1]

The theory that the outline of the terrestrial plate is formed by the uppermost surface of the more collective stratum of the earth's crust, rather than by the actual surface of the earth, appears to have originated from the fact that lightning rods and the buildings to which they were fixed, have occasionally been struck when the earth connections of the rods have been sunk in dry ground.[2]

It has apparently been reasoned from this that the falling lightning, having been denied an outlet in moist soil, has reacted injuriously on the rod and the building, and that it is imperatively necessary, in order that an easy passage to the earth's great reservoir of electricity may be afforded to the lightning, that the earth connections of all lightning rods should be rooted in the more moist ground, which is presumed generally to underlie the dry stratum nearest the surface.[3]

This conception of the cause of the injury to the rods and buildings in question appears to rest on the idea of lightning striking the earth from above, and to stand or fall with that idea;[4] but we hope to show, later on, that a much more simple and natural cause can be produced to explain the fact of injuries to lightning rods that are not in good connection with the earth.[5]

On the whole, then, the theory we have been considering appears to have no good foundation, and does not prevent the natural and *primâ facie* view of the question being taken.

This view is clearly that the uppermost surface of the globe, however such surface may be formed, and whether by natural or by artificial substances, constitutes the outline of the terrestrial plate.[6]

[1] II. D 44.
[2] II. D 47.
[3] II. D 21, 22, 49, 54.   II. G 48, 52.
[4] VI. B *b*.
[5] VII. B *f*.   VIII. B *a*.
[6] I. A 24; C 46, 51.   V. A 26—29.

I

VI. D.

This theory will, it is submitted, be found to harmonize with experience regarding the action of lightning; and it is difficult to see what other hypothesis as to the outline of the terrestrial plate would do so.

(D.) THE INFLUENCES OF THE MATERIALS COMPOSING THE TERRESTRIAL CONDENSER.

Having now discussed the functions fulfilled by the earth and by its surface in the economy of the terrestrial condenser, it will be convenient to deal with the relative powers of the various substances, natural and artificial, forming the principal constituents of this surface (and of the condenser generally) that tend to influence the collection of charge and the occurrence of discharge.[1]

The following table, compiled from the works of eminent authorities, and as the result of general research, gives the more important substances in what is deemed to be the approximate order of their relative influences;[2] it must not, however, be considered as other than an imperfect one, for there is, except as regards metals, very little accurate information in existence on the subject.

Substances which are not usually in a position to affect the terrestrial condenser in any important degree, in regard to life or property, are purposely omitted.

[1] V. A b; B 7—9.   [2] I. A 26; B 1—16.

VI. D a.   (a) TABLE OF INFLUENCES OF VARIOUS SUBSTANCES.   171

Group.	Characteristics of influences of substances in group.			Relative order in influence.	Substances.	Some of the usual forms of occurrence in connection with the terrestrial condenser.	Approximate proportional influence of pure substance.
	When forming portions of collecting or condensing plates. Quick Collectors of Charge.	When forming portions of dielectrics. Great Facilitators to Explosion.	When in a position to transmit current. Good Conductors.				
I. (Metals.)				1st	Silver	Watches, watch chains, money (in purses), alloy in points of lightning rods	1,428 ([1])
				2nd	Copper (including Bronze and Brass)	Lightning rods, roof coverings, nails of slate roofs, balls, vanes, weathercocks, bells, bell wires, bedsteads, turret clocks, telegraph cables, house boilers, statues, portions of dress	1,427 ([1])
				3rd	Gold	Watches, watch chains, money (in purses), personal ornaments, gilding on the following articles : picture frames, wall-paper, furniture, organ pipes, dials of turret clocks, finials, balls, vanes, weathercocks, and crosses	1,143 ([1])
				4th	Zinc	Roof coverings, galvanised iron articles, chimney pots and cowls, baths, speaking tubes	414 ([1])
				5th	Platina	Points of lightning rods	257 ([1])
				6th	Iron	Corrugated iron buildings and roof coverings, chimney pots and cowls, ventilators, roof frames, floor joists and girders, columns, ornamental ridges or crests to roofs, finials, spindles, crosses, window bars, shutters, balustrades, balconies, verandahs, railings, gates, fastenings, fences, rainwater-pipes, eaves gutters, water service pipes, water mains, gas-pipes, gas mains, gaseliers, gas standards, gas meters, wire coverings to stained windows, cramps and dowels in masonry, hoop-iron bond, tanks and cisterns, conservatories, grates and ranges, sash weights, telegraph wires, lightning rods, hot-water pipes, drain pipes, staircases, baluster bars,	240 ([1])

(*Continued.*)

([1]) I. B 12.

172

TABLE OF INFLUENCES OF VARIOUS SUBSTANCES.—(*Continued.*)

Group	Characteristics of influences of substances in group	Relative order in influence	Substances	Some of the usual forms of occurrence in connection with the terrestrial condenser	Approximate proportional influence of pure substance
VI. D a. I. (con.) Conductors.	When forming portions of collecting or condensing plates. Quick Collectors of Charge (*continued*). / When forming portions of dielectrics. Great Facilitators to Explosion (*continued*). / When in a position to transmit current. Good Conductors (*continued*).	6th	Iron (*continued*)	bridges, piers, railways, engines, pumps, boilers, ships, gas holders, oil tanks, fortifications, guns, muskets, fowling-pieces, swords, tires of wheels, agricultural implements, bedsteads, articles of dress	118 (¹)
		7th	Lead	Ridges, hips, valleys, and flashings to slate roofs, eaves gutters, rain-pipes, roof coverings, beds of close-fitting stonework, diamond work and tracery for stained glass, gas-pipe joints, water service pipes	23 (¹)
		8th	Mercury	Backs of mirrors	1 (¹)
		9th	Carbonaceous substances	Soot in chimneys, charcoal embers, coke, coal ashes, graphite	—
		10th	Flame	House fires, furnaces, flames of gas burners. [*See.* II. C. 40.]	—
		11th	Smoke	From house chimneys and furnace shafts, atmospheres of large cities. [*See.* II. C 40.]	1 (²) / 2,221
		12th	The Sea	.	1 (³) / 31,246
		13th	Spring Water	Rivers, streams, wells, springs, ponds, lakes	—
		14th	Rain	.	—
		15th	Snow	.	—
		16th	Hail and Ice	.	—

(¹) I. B 12.   (²) I. B 10.   (³) I. B 16.

173

	17th	Living Animals	Human beings (especially when in the open air), horses, donkeys, cows, cattle, sheep, dogs, pigs, &c.	—
	18th	Vegetation	Trees, bushes, hedges, standing corn, grass, new-mown hay, crops of all kinds, foliage generally	—
	19th	Wood	Houses, outhouses, sheds, huts, barns, fences, rafters and boardings of roofs, floors, doors, shutters, windows, staircases, eaves gutters, flagstaffs, boats, ships, ships' masts, spires, belfries, furniture, bedsteads, walking sticks, fishing rods, carriages and carts, verandahs, bridges, piers, telegraph posts	—
	20th	Aqueous Vapour	Clouds, fog, mist	—
	21st	Moist Earth	Ordinary surface soil in England (more or less throughout the year), telegraphic "good earth"	—
III. Bad Conductors. / Slight Facilitators to Explosion. / Slow Collectors of Charge.	22nd	Dry Earth	Surface of ground in countries from which rainfall is periodically or habitually absent, sand, shingle, telegraphic "bad earth"	—
	23rd	Rock and Stone	Rocky surfaces, chalk, limestone, sandstone, granite, building and paving stones generally, slate, concrete, lime, mortar, plaster	—
	24th	Clay manufactures	Bricks, tiles, earthenware, terra-cotta, cement, porcelain	—
	25th	Dried Vegetable substances	Straw (thatched roofs), matting, cotton, linen, canvas, paper, hemp, linseed oil (paint)	—
	26th	Dried Animal substances	Fur, hair, wool (cloth), felt, silk, leather, bone, ivory	—
IV. Non-Conductors. / Restrainers to Explosion. / Insulators of Charge.	27th	Air		$\dfrac{1,417,000,000,000,000,000}{1}$ [1]
	28th	Bituminous matter	Asphalte, tar, pitch	$\dfrac{2,551,000,000,000,000,000}{1}$ [1]
	29th	Glass	Railway stations, conservatories, skylights, windows, doors, shop fronts	$\dfrac{2,692,000,000,000,000,000}{1}$ [1]

VI. D b; E a.

(b) *Remarks on the Table.*

Some kinds of trees and woods probably belong to Group III.

There appears to be very little precise information obtainable as to the relative influences of the building materials, comprised under the heads of "Wood," "Dry Earth," "Rock and Stone," "Clay Manufactures," and "Dried Vegetable Substances."

In fact, there is undoubtedly scope for useful experiment in determining accurately the relative influences of the whole of the substances in Groups II. and III.; for it would manifestly be advisable to know the particular materials best adapted for the construction of buildings[1] needing special protection against lightning strokes, and also the kinds of soil best suited as sites for such buildings.[2]

As regards the influence of the metals, it can be deduced from the table that iron is 240 times more collective of charge and facilitative to explosion than the most collective non-metallic substance,—533,000 times more so than sea water,—7,500,000 times more so than spring water, rain, snow, hail, ice, human beings, animals, vegetation, wood, clouds, fog, earth, sand, shingle, rock, stone, brick, earthenware, straw, cotton, linen, paper, hemp, wool, silk, leather, and bone,—and 340,000,000,000,000,000,000 times more so than air, asphalte, and glass; and we gain from these figures some idea of the enormous power possessed by metals relatively to non-metallic substances for influencing lightning discharges.[3]

(E.) THE DISCHARGE OF THE TERRESTRIAL CONDENSER.

(a) *The various forms of Terrestrial Electric Discharge.*

We now come to the vital question of the discharge of the terrestrial condenser;[4] and so far as our present

---

[1] II. E 20, 21; G 43.   [2] VII. A.   [3] I. E 1—12.
[4] V. A 12—15, 30—32; B 10—12, 21.

VI. E a.

state of knowledge enables us to judge, the following are the forms in which it occurs, viz. :—

1. A *thunderbolt*, or an explosion between the earth and the clouds, through the intermediate air, constituting an explosion of the terrestrial condenser.[1]

2. A *cloud explosion*, or an explosion of a condenser formed by two separate clouds, through the intervening air.[2]

3. A *terrestrial return stroke*, or a discharge back into the earth from its surface, induced either by a thunderbolt or by a cloud explosion.[3]

4. A *terrestrial leak*, or an escape of electricity through the air, between the earth and the clouds.[4]

It is with the thunderbolt that we are mainly concerned, but the actions of the other three forms of discharge are so closely connected with it that their study cannot well be dissociated from it; and they are of great value in assisting us to form a true idea of its exact nature, and of the best means of preventing it.

We will now refer to the principal manifestations by which these various forms of discharge display their presence.

We know that both thunderbolts and cloud explosions are generally manifested by what are known as thunder and lightning;[5] return strokes are necessarily without atmospheric manifestation, and show themselves chiefly in the forms of shocks and currents;[6] whilst leaks are occasionally visible in the form of auroræ,[7] heat or sheet lightnings,[8] and St. Elmo's fires.[9]

On examining the conditions of these phenomena, it will be found that the element essential to the luminous appearances connected with them is the air.

[1] I. D 1, 2, 3, 11.   V. A 13, 30; B 10, 11, 21. VI. A *a*.
[2] I. D 16.   V. A 30.   [3] I. D 53.   V. A 14, 31; B *e*.
[4] I. A 16.   II. C 11.   V. A 15, 32; B 15—20, 21.
[5] I. D 16.   [6] I. D 53, 66—70.
[7] I. C 78.   [8] I. D 4, 20.   [9] I. C *h*.

VI. E b.

The visible token in the air of explosions is lightning, and of leaks, light.

We are taught that lightning is the appearance of incandescent matter suspended in the air.[1] It cannot, therefore, be the proper term with which to describe the explosion itself, nor can it be correct to apply the term to that portion of a thunderbolt explosion which passes through any other substance than air, *e.g.* through walls, metals, or human beings.

In England the word lightning is generally used to express promiscuously thunderbolt explosions, cloud explosions, leaks shown by "heat lightnings," and the luminous phenomena attendant on these discharges.

In France, however, the term *la foudre*, or thunderbolt, is always used in designating a lightning discharge with the earth,[2] whilst *l'éclair* expresses the lightning itself.

The term *thunderbolt* has the confirmation of lexicology and the sanction of antiquity; and it evidently expresses conveniently the distinction between the harmless lightning that plays among the clouds, and the terrible shafts that visit the earth.[3]

(b) *The Rationale of Thunderbolts.*

Explosion is evidence of work,[4] and, as we have seen, work is, electrically speaking, the product of quantity and potential.[5]

There may, however, be, on the one hand, immense quantity present, and yet nowhere sufficient potential to determine explosion; and, on the other hand, an enormous potential may be developed, yet with such minute quantity that again the combination is inadequate to produce explosion.

In the latter case, however, discharge might probably ensue in another form, viz. that of a leak.[6]

---

[1] I. D 5, 6.   [2] I. D 3, 11, 18.
[3] I. D 1, 2.   [4] V. A 12.
[5] V. B *a*.   [6] II. C 11.   V. A 32; B 17, 18.

VI. E b.

A terrestrial explosion, or thunderbolt, requires, therefore, for its formation two distinct agencies, viz. sufficient quantity, and sufficient potential; and it only occurs at a point on the earth's surface when the combined effect of the quantity and potential accumulated thereat, and of the reciprocal quantity and potential accumulated on the under-surface of the clouds, is powerful enough to overcome explosively the restraint of the intervening air.[1]

The quantity originates with the charge from unknown causes acting from below, and the high potential is due to the condensing influence of the clouds.[2]

It is clear that the immediate cause of all discharge must be charge. It is therefore to charge, and to all circumstances that tend to collect it, and to raise its potential, that we must first direct our attention in investigating the origin of thunderbolts.

Where lightning discharge is seen at the earth's surface, there, it is obvious, that charge must, just before, have existed.[3] To put the same fact in another form, *it is only at the spot where the charge that causes the lightning resides that the latter can possibly "make earth."*

This fact appears to have been constantly, almost systematically, lost sight of; but it must undoubtedly be at the root of all inquiry as to the action of lightning on the earth, and as to the best means of defending life and property from its effects.

Of course, if we knew the original cause for the existence in the earth of electricity, and the exact method in which it collects itself at places on the surface, we should have the best, and indeed only sure, foundation for our endeavours to prevent thunderbolt explosions; but, as we have already said, we know at present next to nothing of this cause and method;[4] so all that we can do is to make the

---

[1] I. D 11, 17. V. B 11.   [2] I. C 36.   [3] V. B 11.
[4] VI. A b.

VI. E b.

most use possible of the facts and laws to be deduced from the researches and experiments made by eminent men,[1] and to study the experience which, unfortunately, is continually accruing from the very disasters which we wish to prevent.[2]

Having settled then that the existence of charge or electricity on the surface of the globe is the cause of lightning explosion, and being unable to account for the origin of this electricity, our investigations must be essentially devoted to the conditions that tend to affect its explosiveness.[3]

[1] I. 2—16, 19, 21—24, 28, 31, 32—35.   II. C 26.
[2] I. G 1—42.   III. 1—203.   [3] V. B 12.

CHAPTER VII.—THE ACTION OF THUNDERBOLTS.

(A) The Electrical Conditions of the Earth's Surface.

(*a*) *The Accumulation of Electricity on the Earth's Surface.*

Although, as we have said, both quantity and potential are necessary in order to produce explosion, still the main and immediate element in its production is necessarily potential;[1] it is therefore to the circumstances of the terrestrial condenser affecting the accumulation of potential, that we must principally direct our attention; and, first, we have to consider the conditions of the natural surface of the earth in this respect.

The earth collecting plate receives its charge, by a method unknown to us, from the source in the interior.

This charging process goes on for a certain period of time, and presumably with some uniformity of action.

The area thus charged, limited probably by the geological conditions of the earth's crust below it, usually consists of surfaces composed of substances of various degrees of influence.[2]

The quicker collecting of these substances would, *cæteris paribus*, collect their full measure of any limited quantity of electricity with which the whole area might simultaneously commence to be charged, in a proportionally shorter time than would the substances of lesser collectivity.[3]

VII. A b.

Hence, if the area should be charged from a source of unlimited electricity, the more collective substances or surfaces would, in any given time whilst the charging process lasted, collect the greater charge, *i.e.* would (capacity being unaltered) attain to a higher potential.[1]

Now the earth is undoubtedly a practically unlimited source of electricity,[2] and it may be presumed that, just previous to a thunderstorm, its surface is charged in such a manner that electricity is continuously accruing thereon, though perhaps at a rate almost imperceptible.

It follows then that the more collective portions of this surface will collect, in any given time during this charging, the more electricity, and thus will obtain the higher potential.

The various portions of the charged area being, previous to a thunderstorm, in these relative states, would all accumulate potential at a much greater rate as soon as a thundercloud should begin to condense them;[3] but these accelerated rates would again vary (on the principle just enunciated) in proportion to the different collective powers of the surfaces.

The result is, therefore, that the more collective any particular portion of the earth's surface may be, the more will potential tend to accumulate thereat, and the more likely will explosions be to spring therefrom.[4]

(b) *Surfaces of Water.*

On referring to the Table of Influence, it will be seen that water of all kinds is the most highly placed of all natural collectors on the earth's surface, and that the sea takes the precedence.[5]

This points to the danger that all bodies on or near

---

[1] V. B 1, 7, 12.   [2] I. C 52, 53.   [3] V. A 28.
[4] I. C 37—45; E 6; G 31—34, 39. II. D 54; G 28. VI. D *a*. VII. A *h*.   [5] VI. D *a*.

VII. A *b*.

natural sheets of water, and especially ships on the sea, are subject to during thunderstorms.[1]

An additional element of risk would appear to attend on vessels when they are in motion during such storms, since it would evidently be possible for a vessel to proceed from one explosively charged area to another, or to keep company, more or less, with a highly condensing cloud in its course above the ocean, either of which circumstances would tend to cause the ship to be the scene of repeated explosions.[2]

There is, however, a very important aspect of large collective surfaces to be noticed, viz. the capacity of their dielectrics, *quâ* area; for the action of this area would, proportionally to its extent, reduce the probability of potential accumulating at any particular portion of it.[3]

The immense uniformly collective area of the ocean would, therefore, by increasing the capacity tend to decrease the potential, of any charge arising at its surface from below, and this would constitute a source of protection to ships; though, on the other hand, it would give scope for the presence of a charge in greater quantity.

Sheets of water, such as small rivers, streams, pools, and ponds, would probably be the natural surfaces of all others that would most assist in bringing about explosion, provided their areas were not too limited to furnish a sufficient quantity of electricity; and, as a matter of fact, we know that when lightning does occur between the earth and the clouds, it generally "makes earth," or appears when close to the earth, at pools and places where moisture abounds;[4] and this (as we have before submitted) is only another way of expressing the fact that the charges which caused the lightning accumulated at these places, and that from them the lightning sprang.

---

[1] II. E 1. VII. C 2.   [2] VII. C 11.   [3] V. B 1, 7, 12.
[4] I. C 37—39, 43, 45; E 6; G 39. II. D 43, 49, 54; G 48. VII. A *h*.

VII. A c d.

(c) *Moist Earth.*

Moist earth, or telegraphic "good earth," such as constitutes the greater portion of the surface soil of England in its ordinary state, is clearly, according to the views we have advanced, somewhat receptive of charge.[1]

It becomes evident then that the best positions for erecting buildings, so far as their defence from the action of thunderbolts is concerned, are away from the banks of rivers and lakes, and from the vicinity of pools, streams, and moisture generally; that the better drained and the drier the ground is around the buildings, the better; and that the theory of moist earth being necessary in the neighbourhood of a building for the purpose of defending it from the effects of lightning is exactly contrary to the real requirements of the case.[2]

Vegetation, which *per se* is more collective than moist earth, probably increases the collectivity of the ground;[3] and this idea accords with the apparent fact of the greater frequency with which thunderbolts occur in the fields and in the open country than in the towns.[4]

As in the case of the sea, the potential of the charge on moist earth would tend to vary inversely as the extent of collective area over which it was spread, so that a comparatively small portion of moist earth circumscribed by rock or dry earth would be all the more dangerous.[5]

(d) *Rocky and Dry Surfaces.*

From what has been stated, it is clear that rock and dry earth, being less collective than ordinary moist earth, would, in any given time, collect a smaller charge, *i.e.* would not attain so high a potential.[6] Hence, such surfaces would tend to be sources of protection, and a house built on a rocky or very dry surface, or at a distance from

---

[1] VI. D *a*.   [2] II. D 7, 22, 26, 28, 36, 43, 49, 53, 54; G 52.
[3] VI. D *a*.   [4] II. G 19—21, 35, 36.   [5] VII. A *h*.
[6] VI. D *a*. VII. A *h*.

VII. A *e.*

moisture, ought to be, *ipso facto,* less liable to be struck by lightning.

This view is fortified by the experience, already cited, that lightning is found, as a rule, to seek moisture;[1] for this clearly implies that it avoids, by preference, places where moisture does not abound, viz. rocky and dry surfaces.

By rocky surfaces we mean those formed of the bare rock, and not merely rocky sites; for it will frequently be found that rocky sites are more or less overlaid in places by thin coverings of soil or sod, and places so covered would not come within the category of rocky surfaces.[2]

Even coverings of snow and ice on otherwise rocky surfaces, as *e.g.* Alpine summits, would doubtless completely alter their character, and would make them more susceptible of collecting charge than the bare rock; and there is reason to believe that, in most cases where explosions spring from rocky sites, these sites have been covered with some extraneous substance tending to faciliate the accumulation of electricity.[3]

The curious "fulgurites" that is occasionally produced where the ground is struck by lightning is an apparent instance of explosion occurring from a slowly collective surface; but without a full knowledge of all the concomitant circumstances in cases of "fulgurites," it would be difficult to form an accurate opinion as to its cause.[4]

On the same principle as that already mentioned in regard to collective surfaces, rocky, dry, and all slowly collective or insulative surfaces would be proportionally less influenced by the condensing action of thunderclouds.

(*e*) *Paved Surfaces.*

From the consideration of surfaces naturally rocky we are led to that of surfaces artificially so, viz. stone pavements.[5]

VII. A *e*.

Here we appear to have very similar conditions of slow collectivity, though probably the thickness of the paving would be an element in the question, since it cannot be supposed that, in practice, however theoretically correct the idea may be, a mere film of slowly collecting matter on the earth's surface would have so much effect in preventing explosion as a thicker stratum; for since the surface receives its original charge from below, the thicker the mass of slowly collecting substance immediately below the surface should be, the farther would the bulk of this charge be kept from it, and the less likelihood would there be of its receiving any appreciable quantity of electricity.

In the case of ordinary well-laid stone flagging, there would seem to be good ground for presuming that the surface would be proportionally less collective than if the soil had not been covered at all.

Brick pavements would probably be more insulative than stone, and asphalted ones more so than those of brick.

On the same principle we have reason to expect that a layer of metal laid on the earth's surface could greatly increase the collectivity of the plate thereat.

It is submitted that the paved surfaces abounding in cities and towns, and especially close around buildings therein, are among the causes that contribute to protect the buildings from thunderbolts.[1]

The essential idea that we have been urging regarding the influence of rocky and slowly collecting surfaces is that the electricity derivable from the interior of the earth has every reason to accumulate by preference in places of good collectivity, and that even if slowly collecting surfaces do become charged from below, they are only in a comparatively small degree condensed by the clouds; hence such surfaces are the less likely to accumulate potential, and consequently explosions are the less likely to spring from them.

---

[1] I. G 11, 35, 19—21.

VII. A *f g.*

(*f*) *Surfaces formed by Railway Metals.*

An artificial feature, partaking somewhat of the nature of metal pavement, and appertaining largely to countries where civilisation prevails, is the network of iron, formed by the railway system.[1]

Here we have a surface highly capable of accumulating potential, *quâ* its metal,[2] but of great capacity *quâ* its surface,[3] and thus in a condition somewhat analogous to that of the ocean.

The influence of railways cannot, however, be thoroughly discussed without taking into account a third condition attached to them, viz. the leaks occasioned by the angularities of their metals. The subject will therefore be dealt with again under the head of Terrestrial Leakage.[4]

(*g*) *The Shape and Geological Formation of the Ground.*

Turning now to the topographical feature of the ground,[5] it is evident that the higher any place is above the level of the sea, and, hence, the nearer to the clouds, the less thick is its dielectric.

General elevation thus tends to reduce capacity, *quâ* its thickness, and therefore, *cæteris paribus*, to increase potential; hence it is a source of danger.[6]

Table lands, mountain ranges, and watersheds generally, would on this account be regions in which potential would tend to accumulate, in preference to basins.

On the same principle, any feature of the ground elevated above its surroundings, independently of its general level, would be, *quâ* that fact, a source of comparative danger, and the summits of mountains and hills

---

[1] VII. A *h.* [2] VI. D *a.* [3] V. B 1, 7, 12.
[4] VII. G *d.* [5] VII. A *h.*
[6] I. D 17, 35. II. C 43; G 26. V. B 1, 7, 12.

VII. A *h*.

would be places more exposed to risk than the sides and valleys.¹

Lateral prominence of ground would also lead to increased exposure to chances of condensation by a cloud approaching, but not as yet arrived at, the zenith of the place, in comparison with other portions of ground not horizontally exposed in so great a degree.

The sides of hills and the sea coast are examples of this condition of prominence.

The geological nature of the earth's crust at any place would probably constitute a most important element affecting the occurrence of charge on the surface.²

So little, however, appears to be known of the relative influences of the various kinds of rocks and geological formations as regards terrestrial electricity, that there is not much scope for enlargement on the subject;³ though there is probably not much doubt that certain formations favour the collection of electricity much more than do others.

A thorough study of the earth's crust, with reference to terrestrial electricity, and to the actual localities of thunderbolt explosions, would probably throw much light on many points as to which we are at present in darkness, and would materially assist in solving the problem of the cause of the earth's electricity.

(*h*) *Analysis of Incidents in regard to Conditions of Surface.*

The following incidents in Chapter III. contain more or less allusion to conditions of surface:—

Sea (except cases of Ships).
Nos. 44, 83, 96, 117, 123, 179, 180.

Rivers.
Nos. 15, 125, 127, 137, 181.

---

¹ I. G 33, 40, 41.   ² I. C 41; G 30; VII. A *h*.   ³ VI. D *b*.

Lakes.
Nos. 30, 31, 85.

Wells, Pools, and Streams.
Nos. 49, 82, 103, 155.

Dry Earth, Sand, or Rock.

Nos. 22,	86,	117,	156,
44,	87,	128,	157,
54,	89,	131,	178,
55,	91,	147,	179.
64,	103,	155,	

Other natures of Soil or Ground.

Nos. 10,	117,	168,
54,	118,	176,
56,	129,	182,
90,	147,	202,
116,	161,	203.

Charcoal Trenches.
No. 47.

Pavements.
Nos. 24, 112, 181, 183, 192, 203.

Railways.
Nos. 88, 98, 151, 198.

Elevated or Exposed Ground.[1]

Nos. 4,	95,	117,	155,
55,	108,	128,	156,
79,	116,	147,	181.
94,			

Valleys or Low Ground.
Nos. 3, 54, 59, 203.

(B) DETAILS OF THUNDERBOLT ACTION.

(a) *Classification of objects on the Earth's Surface.*

We now come to the consideration of the principal objects, not forming integral portions of the earth, met

[1] VII. B e.

VII. B a.

with on its surface, such as the artificial features presented by buildings, ships, and constructions of all kinds, and the extraneous features formed by natural objects, such as human beings, animals, and trees.

In Chapter VI., Section C, we have submitted that the uppermost surface of the globe, whether such surface is formed by natural or by artificial features, constitutes the outline of the terrestrial plate.

Under this aspect, it is evident that the sides and upper surfaces of the artificial and extraneous objects we are now dealing with, form, wherever these are isolated features on the natural surface of the earth, portions of the earth's collecting plate.

An important element, however, comes into play with these objects, and that is their amount of electrical connection with the earth's natural surface.

If this connection be good, then the object under consideration, *quâ* its exterior sides and surfaces, takes an active share in the general collecting plate, and constitutes what may, perhaps, conveniently be called a local collecting plate, or *local plate*.

If, however, this electrical connection should not exist, or should be bad, then the outer surfaces of the object, whatever may be their collectivity, are, like the rocky and dry surfaces considered in the last section, merely passive portions of the general collecting plate; but, unlike the rocky surfaces, the objects we are now treating of present, as a rule, when taking only a passive share in the earth's collecting plate, a special feature which causes them to play an active part in another sphere of influence.

This special feature consists in the more or less vertical surfaces presented by the sides of buildings and other objects, which surfaces must materially influence, according to their powers of restraint, the dielectric immediately over the portions of the natural surface of the earth lying close outside the bases of these vertical sides.[1]

[1] V. A 23, 29.

VII. B *b*.

In fact, these sides necessarily form *local dielectrics* to the ground immediately adjacent to them, and the whole of the outer surfaces of the building or object, so long as their direction with regard to the ground in question is at all inclined upward, must, since the direction of a condensing cloud may be at almost any angle above a horizontal plane through any part of the building, take a greater or less share in these local dielectrics according to the circumstances of the case; and indeed, an instance is on record of a thundercloud having been even below the level of a building struck through its agency.[1]

All the objects, then, to be met with on the earth's natural surface, are capable of being grouped either as local plates or as local dielectrics, according as they are, or are not, electrically connected to the ground.[2]

(*b*)　*Electrical Connection.*

It becomes necessary now to consider what constitutes good electrical connection.

Good electrical connection between two articles or substances means the existence of such intimate contact between them, that electricity freely distributes itself between the two, without meeting obstruction at the points of contact.

This condition can only apply between two articles or substances which are themselves good collectors or conductors, and cannot be said to exist in the case where either body is a slow collector, still less where one of them is an insulator.[3]

Electrical connection is not the same as mechanical contact, for even when the latter exists between two articles of the same metal, considerable hindrance to electrical distribution is frequently caused by the slight film of air existing at the apparent contact; and to make

[1] III. 4.　　[2] VII. C 15—18.　　[3] VI. D *a*.

VII. B c.

good electrical connection between an object and the ground, the presence of some degree of moisture appears to be generally requisite, the action of which seems, by means of some kind of electrolytic action, to render contact more electrically perfect.

For a collector, then, to be in good electrical connection with the earth, it would generally be necessary that the surface crust should be more or less in a moist state, and that the collector should either be itself in close contact with this crust, or should be joined thereto by some form of continuous metal, this metal being itself incorporated with the collector at one end, and buried in the ground at the other.

No other substance than metal would seem to provide efficient connection between the earth and a metal or collector separated from it.

It is then assumed, as a broad rule, in the following paragraphs, that an object not composed of collective material, or, if a collector, not in contact with the surface of the earth nor joined thereto by metal, is not electrically connected to the ground, and consequently forms a local dielectric; whilst those collectors that are thus electrically connected constitute local plates; and by the term "ground" or "earth," the kind of ground, in its average state of moisture, which forms the ordinary surface soil of England, is intended.[1]

(c) *Explosive Action.*

Explosive action necessarily follows the line of least restraint between the points of explosion on the collecting and condensing plates.[2]

Experience shows, however, that in the terrestrial condenser it is quite impossible to foresee the exact direction and path of this line; and this is the less remarkable when we consider that the plate formed by the cloud is

---

[1] VI. D a. [2] I. D 31, 32. V. A 30.

VII. B c.

always more or less in motion, and that the precise position it will occupy over any particular building or place on the earth's surface at the moment when potential has accumulated to explosive point can obviously never be predicted.[1]

The zigzag appearance of lightning is an undoubted proof of the irregularity of the course of an explosion's line of least restraint. Explosive action can only exist in a restrainer or slight facilitator;[2] and it always proceeds *through* one of these substances, though this piercing action may occasionally take a path coinciding with the plane of contact between two dissimilar substances, as *e.g.* over the exterior surface of a wall whilst piercing the film of air in contact therewith.[3]

The fundamental law concerning explosive action appears to be that it springs *from* the collecting and condensing plates, and acts *within* the dielectric.

Explosion, therefore, affects terrestrial objects very differently, according as they form local plates or local dielectrics.

The chief manifestations of terrestrial explosive action appear to be as follows, viz. :—

        1. Spark or Lightning.
        2. Heat.
        3. Expansive or rending force.
        4. Uplifting force.
        5. Shock to animal systems.[4]

Explosion, when passing through the air, is manifested by lightning and heat; and, when through other restrainers, or through slight facilitators, by rending, uplifting, and heat.

In the case of local plates, explosion only injures them in the act of springing from or *leaving* them, and the

---

[1] II. E 35.   [2] V. A 22, 23. VI. D *a*.
[3] II. G 33.   [4] I. D 16, 37, 39, 40, 41, 44—46, 53.

VII. B c.

injury takes different forms according to the substance of the plate.[1]

Thus when metals form local plates, explosion frequently fuses them at the exact places where it leaves them, where, in fact, it is in contact with them in the form of great heat, viz. lightning; but it does not otherwise damage them.[2]

In leaving human beings and animals it usually causes a fatal shock to the system, accompanied by traces of burning.[3]

In leaving woodwork, it generally shatters the extremities, and occasionally sets it on fire.[4]

In its action on local dielectrics, explosion is necessarily more violent, and is accompanied by an uplifting force or blow.

When it passes through metals forming local dielectrics, it generally fuses them at the points where it enters, as well as where it leaves them, at which points the character of the explosion is necessarily influenced by the non-metallic substances which it leaves and enters, respectively; but the bodies of the metals, except in the case of thin wires, do not appear to be heated or otherwise materially affected, a circumstance probably due to the great influence possessed by metals for facilitating the passage of the explosion when they are strong enough to resist its mechanical shock; in other words, to the fact that they do not afford sufficient time for the work of an explosion passing through them to manifest itself in the form of heat.[5]

Thin wires are occasionally recorded as having been melted, such having doubtless at the time been considered as the most obvious manner of accounting for their disappearance; but it would seem probable that the metal of very small wires, *e.g.* bell wires, is of too small extent to exercise much facilitating influence on the explosion, and that they are usually disintegrated and dispersed by its

---

[1] VI. D a. [2] I. D 36, 47. VII. C 3. VIII. C 9.
[3] VII. C 8, 9. [4] VII. C 1 μ, 6.
[5] I. A 31; D 36, 47. VII. C 3. III. 27, 28.

VII. B c.

mechanical force or blow,[1] and the not infrequent action of explosion in breaking in pieces slender lightning rods not electrically connected to the ground, would appear to confirm this view.[2]

When human beings form local dielectrics, they must experience, firstly, the blow from the explosion; secondly, the shock due to its passage through their bodies; and thirdly, the shock due from its contact on leaving them.[3]

When woodwork forms part of a local dielectric, explosion appears sometimes to shiver it, and at other times to set it on fire.[4]

Trees and woods probably differ from each other considerably in their influence, and little seems to be known as to their relative powers for collecting charge or for facilitating explosions. A portion of a construction made of one kind of wood might act as a plate, whilst the same object in the same position, made of another kind, might act as a dielectric.[5]

It is when passing through brick and stone, and through all restrainers and slight facilitators, that an explosion manifests itself most violently. It here exhibits a powerful rending force,[6] which necessarily has a disastrous effect on the buildings which experience it, and also an uplifting force.[7]

This latter force occasionally lifts heavy stones, and transports them to considerable distances; and the result of the two forces is frequently not unlike that due to a gunpowder explosion.

Church spires of stone, surmounted by metal work, and containing within them much metal in different forms at various elevated levels, are more especially apt to experience the rending force of thunderbolts.[8]

When it is merely the exterior surface of a brick or

---

[1] II. B 22.   III. 18, 19, 52, 65, 67, 119.   [2] VIII. C 5, 10.
[3] VII. C 8.   [4] VII. C 1 μ, 6.   [5] II. G 33, 39.   VI. D a, b.
[6] I. D 45, 46.   VII. C 1 π, 14.   [7] I. D 40.   VI. B a.
[8] VII. C 1, a, δ.

VII. B d.

stone building that acts as a local dielectric, it is probable that there is less scope for the display of this rending action, since most of the expansive force would be dissipated in the atmosphere.

Explosion occasionally ploughs a path horizontally along the surface of the earth, for some distance from the place whence it springs, before it utilises an object on the surface in order to rise therefrom.[1]

In our present state of knowledge it would probably be quite impossible to give exhaustively all the various manners in which explosive action is liable to act; but, from what has been submitted, it would appear certain that the injury it effects on an object is as a rule far greater when that object is a local dielectric than when it is a local plate.

(d) *Local Plates.*

If a construction of wood, or of metal, or of any substance of good collectivity, should be electrically connected to the ground, the exterior surfaces would receive charge therefrom, and would constitute a local plate.

A wooden or metal building, thus circumstanced, being more collective than the surrounding surfaces,[2] would, on the same principle as that already mentioned in the case of two unequally collecting surfaces of ground, attain to a higher potential; for the combined surfaces of such a building would in any given time accumulate a greater amount of electricity than an equal area of ground.

It is thus evident that a metal building is liable to form a local plate of great accumulating power,[3] and that an explosion is, *cæteris paribus*, more likely to spring from such a building than from the ground immediately around it.

All metal constructions and objects, however, when forming local plates, contain the element of leakage in a

VII. B d.

greater or less degree; and this element (which is treated on hereafter in Section F[1]) probably exercises considerable influence in preventing explosions from them.

The comparative absence of leakage conditions in wooden constructions renders it likely that, as local plates, they are more dangerous than metal buildings.[2]

We have now to consider brick and stone buildings containing, on their walls or roofs, metal surfaces, such as roof coverings, eaves gutters, rain-water pipes, finials, vanes, and other articles, all electrically connected to the ground.[3]

These surfaces come into the same category as metal buildings, and constitute local plates of more or less importance in proportion to the surface of metal they display; and there is no reason for assuming that any form they may possess such *e.g.* as long slender rods, bands, tubes, or wire ropes, or that any direction these may take, whether vertical or otherwise, in any way alters their collecting power; although this would, of course, be affected by leakage, on the same principle as for metal buildings.

It is clear that an explosion possesses more power to injure a brick or stone building containing various metal surfaces than one formed wholly of metal.[4]

One element, however, in connection with metals generally, would have a great influence over explosion, viz. their elevation above the ground; and this influence would be present in all local plates.

Elevation would act as follows:—In proportion to the height of a local plate above the general surface of the ground the thickness of the air dielectric would be diminished; thus capacity is to the same amount reduced, and (quantity being unaltered) potential becomes proportionally increased.[5]

---

[1] VII. F c.
[2] II. E 42.   VII. C 1 i.
[3] VII. C 3.
[4] I. E 2, 6, 10, 12.   VII. C 3.
[5] I. D 17, 35.   II. G 26.   V. B 1, 7, 12.

VII. B e.

Elevated metal in the form of a local plate is thus a condition tending greatly to promote explosion.[1]

Ships, when afloat, are well connected to the earth's surface (formed by the sea); thus they generally constitute local plates; and, for the same reasons connected with leakage as mentioned in the cases of wooden and iron buildings, wooden ships, would generally be more dangerous than iron ones.[2]

The following objects would probably also be liable, during thunderstorms, to form local plates, viz.:—

1. Human beings and animals standing on the ground in the open air.[3]
2. Trees.[4]
3. Flagstaffs and telegraph poles springing from the ground.[5]

It is evident that a construction or other object may occasionally form partly a plate and partly a dielectric, as *e.g.* when a metal surface on the lower part of a stone building is electrically connected to the ground; in this case all the rest of the building, especially any part above or adjacent to the metal surface or its connection to the ground, is liable to constitute a local dielectric to the local plates formed by the metal.[6]

Local plates are also always liable to act accidentally as local dielectrics to each other, or to adjacent portions of ground to which they themselves are not electrically connected. This condition will be more fully discussed after the question of local dielectrics has been considered.[7]

(e) *Persons in the Open Air.*

That human beings on the ground, in the open air, when killed by lightning, do usually form local plates seems almost certain from the following facts, viz.:—

---

[1] II. G 43. VII. C 15, 16. VIII. C 6.   [2] VII. C 2.
[3] VI. D *a*. VII. C 8, 9.   [4] I. F 5—8. VII. C 5. I. D 44.
[5] VII. C 6.   [6] VII. C 16.   [7] VII. B *g*.

THE ACTION OF THUNDERBOLTS. 197

VII. B *e*.

(1.) They, under the same circumstances, constantly receive shocks from return strokes; and this would be impossible if at the time they had formed local dielectrics.[1]

(2.) If a person, ordinarily clad, and not standing on an insulating stool, touches a charged conductor in a laboratory, the charge escapes through his body to earth; which shows that the leather soles of boots are not insulators.[2]

It is thus tolerably clear that, as a rule, a person, situated as above described, has been previously charged on some part of the surface of the body by the ground; and that death ensues from the fearful shock given to the delicate nervous organization of the human system by the starting of a thunderbolt from the charged surface.

It appears reasonable to infer that the upper and more vital portions of the body, such as the regions of the head and heart, are the particular parts that fatal explosions usually spring from.[3]

The charge that thus explosively leaves them has accumulated, unconsciously to them, on the actual surface of their bodies; and the explosion pierces its way through the slightly facilitative substances, such as cotton, linen, flannel, wool, cloth, silk, leather, felt, of which their clothes usually consist, in the direction rendered least restraining by the presence of metal in and among their clothes, and especially metal of a highly attractive nature, such as gold, silver, and copper.

These clothes and metals must doubtless be looked upon as local dielectrics, and the existence of the metals is probably an element of vital importance, and constitutes the last drop that overflows the electric bucket.[4]

During thunderous weather in the country, it would

[1] I. D 55, 56, 58. VII. C 8.   [2] VI. D *a*.
[3] I. D 59, 60.   [4] I. E 11; F 15.

VII. B ƒ.

appear to be worth the while of everybody (and perhaps of ladies in particular) to reckon up the amount of metal they carry, in one form or another, about their persons before they go on expeditions to places remote from accessible buildings, whether they be walking, riding, or driving.[1]

Of these occupations, riding would seem to be especially dangerous, owing to the good electrical connections formed by the horse's shoes, the horse's collectivity, and the elevation of the rider.[2]

We find, from the records of fatal thunderbolts, that the traces most frequently left by them correspond closely with the course of action on the part of explosions that has just been suggested. Traces of burning are found on the skin, doubtless where the explosion left it; clothes are torn; and watches more or less fused.[3]

The fatal accident that occurred at Schelthorn, one of the Bernese Alps, on the 21st June, 1865, when a young English lady was the victim of a thunderbolt, is an exceptionally distressing instance of the danger to which persons in the open air are exposed to during thunderous weather.

(ƒ) *Local Dielectrics.*

The walls of buildings constructed of brick, stone, and similar materials, are not electrically connected to the ground; therefore the exterior surfaces of the walls constitute local dielectrics to the ground close outside them.

The exercise of the function of local dielectric does not interfere with the property which all outer walls possess in virtue of being portions of the uppermost surface of the earth, viz. the property of forming parts of the terrestrial collecting plate; but, as already stated, this condition can only apply in a passive form to local dielectrics.

If, then, charge should from some cause collect on the

---

[1] I. F 9, 14.   [2] VII. C δ, η, θ.
[3] III. 15, 35, 113, 161, 162, 165.

VII. B *f.*

surface of the ground adjacent to the walls of a brick or stone building, and if a thundercloud should condense this charge, inasmuch as brick and stone are less restraining than air,[1] the dielectric afforded by the vertical film of air immediately in contact with the outer surfaces of the walls, *i.e.* afforded practically by the outer surfaces themselves, is of less restraining power than the adjacent vertical layers of air not thus in contact with the walls, and hence is apt, in proportion as the wall surfaces coincide with the line of least restraint, to hasten explosion; in other words, restraint being decreased, and quantity unaffected, potential becomes increased.[2]

The same state of things applies to the surfaces of wooden and metal buildings insulated from the ground; but with these, since wood and metal are less restraining than brick and stone,[3] capacity is still more diminished.

In fact, the existence of any sort of building not electrically connected to the ground, and consequently forming a local dielectric, tends, in proportion to the relative restraining power of the outer surfaces of the walls as compared to that of air, to cause discharge close outside the walls; but experience proves that this tendency, in the case of buildings of brick, stone, and similar materials, is not usually *per se* strong enough to determine explosion, and is only dangerous to these buildings when they contain metals on their outer surfaces, or when they present prominently elevated features.

Metal buildings insulated from the ground, and buildings of brick, stone, or wood, containing on the exterior surfaces of their roofs or walls metal not connected to the ground, are unquestionably sources of danger, inasmuch as the relative restraint exercised by the metal as compared to that of air is so much less as to be almost in the category of attracting explosion.[4]

---

[1] VI. D *a.*   [2] V. B 1, 7, 12.   [3] VI. D *a.*
[4] I. E 2, 3, 6—12.   V. B 1, 7, 12.   VI. D *a.*

VII. B ƒ.

Certain special dangers to which buildings formed wholly of metal or wood, and acting as local dielectrics, are subject, are dealt with hereafter.[1]

Elevation is a factor that in local dielectrics is of much influence; for, the higher the wall, or the more elevated the features of the building, above the collecting plate below, the more is the air dielectric supplanted, the more is capacity (*quâ* restraint) diminished, and the more is explosion hastened.[2]

Especially is the height of metal on a local dielectric a matter of great moment; and the explanation of this fact appears to be as follows:—The higher the position occupied by the metal, the greater is the attraction exerted over the earth's electricity towards the half-way point in the air where, through explosion, it meets the electricity of the clouds, and the more influentially is the capacity's restraint lessened.[3]

It follows, from what has been advanced, that towers, spires, domes, belfries, columns, tall chimney-stalks, and all prominently elevated features of brick and stone buildings, are, *per se*, sources of danger; and that this danger is much enhanced by the presence of elevated metal not connected to the ground, *e.g.* roofs, spindles, finials, weathercocks, vanes, crosses, chimney-pots, ornamental ridges, and eaves gutters.[4]

Owing to the great danger occasioned by metal local dielectrics, it might in some instances be of advantage to mitigate this danger by connecting the metals to the ground, thereby converting them into local plates.

The surfaces of glass buildings ought, *quâ* the glass, to act as greater restrainers of explosion than air; and glass houses would thus contain an element of security.[5] The wooden or metal framework of the glass might, however, in many cases, counterbalance its restraining qualities.

---

[1] VII. B *i*.
[2] V. B 1, 7, 12.
[3] I. D 49; E 6. V. B 1, 7, 12. VI. B *b*.
[4] II. E 43. VII. C 3, 16, 17, 18. VIII. C 8.
[5] VI. D *a*.

VII. B *g*.

A building whose roof or walls were coated with pitch or tar would, *quâ* this coating, appear to receive additional security.[1]

The sides of shafts leading to mines may probably be occasionally placed in the same category as the exterior sides of buildings.[2]

It seems quite possible that the ground at the bottom of a deep colliery shaft may sometimes receive charge; in this case the vertical sides of the shaft, especially if they contained woodwork or metal, would constitute more or less efficient local dielectrics, and might therefore promote discharge from the bottom; and such a discharge would doubtless ignite any deadly gas that might be present, and cause one of those dreadful colliery explosions which are so constantly occurring.[3]

The natural features presented by the more or less vertical surfaces of cliffs, precipices, cuttings, quarries, and rocky crags, would occasionally constitute local dielectrics.

The ordinary surface of the earth may also form a local dielectric, as is evidenced by the furrows occasionally ploughed by explosions.[4]

(*g*) *Accidental Dielectrics formed by Local Plates.*

A most important feature of thunderbolt action is the fact that a local plate may accidentally form a local dielectric to some other plate.[5]

A piece of metal forming a local plate on the exterior surface of a building may form an accidental dielectric to another piece of metal forming another local plate on the same building; or such a piece of metal as either of these may form an accidental dielectric to an adjacent, though not immediately contiguous, portion of the earth's surface.

In these two cases the explosion leaps from metal to

---

[1] VI. D *a*. [2] ? I. G 34. [3] I. G 42.
[4] VII. C 13. [5] VII. B *d*.

VII. B *g*.

metal, or from ground to metal, and, in the act, probably causes considerable injury to the building.

These cases of explosions, deviating from what might be considered the orthodox road, have been termed "lateral discharges" when they have occurred in connection with lightning rods, and metallic arrangements (discussed in Chapter VIII.) have been devised in order to remedy such erratic courses.

As regards ships, the fact of their hulls forming local plates would appear not always to cause their masts to exercise the same function; for the lower masts of ships have been rent in a manner that would betoken their condition as dielectrics.

Thus, although the lower mast of a ship is usually electrically connected to the hull, the explosion is liable to spring from a point on the surface of this hull through the air to a particular point on the mast much above the level of the deck, and thus the mast is to all intents and purposes a dielectric.

Hence, a local plate may be so shaped that one of its higher features acts as a dielectric to the lower portion.

Again, a human body acting as a plate, *quâ* the ground on which it stands, may yet form a dielectric, by means of its upper parts, to a patch of charged ground near it, but insulated from the patch on which it stands.

The general law on this subject would appear to be that a portion of one local plate may constitute an accidental dielectric to another portion of the same plate, or to another plate, or to any part of the earth's natural surface to which the original plate is not electrically connected.[1]

All these actions point to the conclusion that an explosion carefully and minutely prepares beforehand its line of least restraint, according to its own views of what constitutes restraint,[2] and quite independently of any channels of conduction that might on other grounds be supposed to lead more directly to the electricity in the clouds.

(h) The Protection afforded by the Interiors of Buildings.

The action of thunderbolts in connection with the exteriors of buildings and other hollow constructions having been considered, the question arises as to how far their interiors are exposed to danger.[1]

Universal experience leaves no doubt that the insides of houses are, as a rule, safe asylums from thunderbolts, and it is obvious that this security must be owing to some function of the roofs and walls.

This function is evidently that due to the fact of the exterior surfaces of all constructions on the earth's surface forming parts of the terrestrial plate, whence it follows that any spaces existing below or within these artificial exterior surfaces are below or beyond this plate, and consequently *outside* of the terrestrial condenser, and hence not subject to the action of its explosions.[2]

Therefore, whether a construction forms a local plate (as *e.g.* a ship afloat or a metal building connected to the ground), or a local dielectric (as *e.g.* a brick or stone building), the mere facts of resting on the natural surface of the globe and of being hollow, constitute to the interior of the construction a source of protection from thunderbolts.[3]

On the same principle, caves and underground constructions would, *quâ* being underneath the terrestrial plate, be places safe from thunderbolts; and it is worth noticing that the ancients were cognisant of this property due to caves, and took advantage of it.[4]

On the same principle also, the ground exactly underneath the walls of buildings cannot, as a rule, collect charge; and the walls are therefore generally in no danger of being actually blown up or rent longitudinally.

VII. B i.

(i) *The Dangers to which Interiors are liable.*

Although, however, we may accept the general rule that the interiors of buildings, of ships, and of hollow constructions generally, are not exposed to thunderbolt action, still there are important exceptions to this rule, and the interiors of constructions forming local dielectrics are frequently subject to danger.[1]

It is evident that the whole substance, the outer surface of which constitutes a local dielectric, may occasionally share to some extent with the outer surface the properties of this dielectric.

Thus, an explosion, instead of confining itself to the exterior surface of a building, may possibly pass within the substance of its walls or of its roof; or it may proceed for some distance over the outside, and then pass inside through an opening; or there may be some great attraction inside which it will pierce the wall in order at once to reach.

The side or top of a hollow construction may therefore possess certain features which, independently of exterior characteristics, as *e.g.* metals and elevation, might effectively influence explosive action, and might determine a line of least restraint in the highest degree dangerous to the inmates of that construction.

We will notice some of what appear to be the principal conditions that induce explosion not to confine its action only to the exterior surfaces of constructions, when these form dielectrics.

In the first place, we must recount what we have already advanced regarding explosive action generally, viz. that it always pierces dielectrics, and that it is the film of air immediately contiguous to the exterior surface of a wall that it pierces when it apparently passes over this surface.[2]

VII. B *i.*

In the case of a brick or stone building this film is, as we have said, of less restraining power than air generally, *e.g.* than air at only a minute distance from the surface; but that the brick and stone itself is only slightly facilitative is proved by the fact that explosion actually passing through or within it always rends it open.[1]

Therefore, and also because the charge causing the explosion is outside the house, and in most cases the line of least restraint therefrom to the condensing cloud, whilst the charge is undergoing condensation, lies also outside the house, explosion does not usually pierce the brick or stone walls, nor otherwise enter the house, unless there are in the walls or interior some special facilities for its progress.

Now these facilities would appear to be of three natures:—

> (1.) There might be openings in the masonry (as there usually are), and during a thunderstorm a window, door, shutter, or other means of closing an aperture may be left open, or the aperture may be permanently open as in some spires, towers, and belfries, and close to it, inside the building, there may be a piece of metal, or a human body, or some other facilitator.
>
> (2.) There may be pieces of metal passing transversely through the walls, as *e.g.* tie-bars, girders, cramps, and hoop iron bond.
>
> (3.) There may be pieces of metal in contact with the inner surfaces of the walls, as *e.g.* water-pipes, gas-pipes, clocks, bells.

In all these cases there would be an inducement to some extent for an explosion to enter a house, though it by no means follows that the inducement will be sufficient; and the last case is the most dangerous of the three, since it necessitates the piercing of the wall.[2]

[1] VII. B *c.* VI. D *a.*
[2] I. E 7, 10. II. G 53. VII. C 3, δ, ϵ, 8, *a*, γ.

VII. B k.

One or more of these conditions would frequently apply to towers, spires, cupolas, domes, belfries, and other elevated features of buildings.

In every case the danger would probably be enhanced if the walls were of unusual thinness.

When we come to consider local dielectrics whose walls or sides are made of metal, wood, or other facilitative materials, there is evidently a greater possibility that the thunderbolt may pass to the inside, since these materials facilitate the passage of the explosion through their substances;[1] and on the whole it is probable that constructions of this nature, *e.g.* houses, or parts thereof, gas-holders, and oil tanks, forming local dielectrics, are always more or less dangerous to their contents,[2] and that wooden houses, and wooden constructions generally, when in this category, being less facilitative and more likely to be set on fire, are more dangerous than metal ones.[3]

(k) *The Special Danger from Chimneys.*

The most frequent source of exception to the security afforded to buildings by their roofs and walls is the danger arising from their chimneys, and this is one of the most fertile sources of injury and death from thunderbolts.[4]

Dwelling-houses in nearly all civilised countries contain some means of warming their interiors, or of cooking the food of their inmates; and this necessitates some means of outlet for the smoke and gases due to combustion.

In this country all habitable houses possess, within the thickness of their walls, one or more nearly vertical chimney shafts or flues.

These shafts extend either from the bottom of the walls or from various levels up their height (corresponding to

VII. B *k.*

the stories of which the house is composed), to some distance above the highest points of the roofs, where they terminate in the prominent, isolated, brick or stone columns formed by chimney-stacks.

These stacks are frequently themselves surmounted by tall chimney-pots, consisting of earthenware or metal cylinders.

At the bottom of each shaft is usually a mass of metal in the shape of a grate or range, which is generally insulated from the ground by the back hearthstone.

The shafts themselves are necessarily lined with soot or other unconsumed products of combustion; and when the fires are lit, there is a mass of flame at the grate, and a column of smoke or gaseous matter continually ascending the shaft and pouring into the air.[1]

In the case of a three-storied country house, there is a kitchen chimney shaft perhaps 80 feet high and 5 or 6 feet in girth, springing from a large range, which includes copper, brass, and iron in its composition.

In the case of a labourer's one-storied cottage, the chimney-shaft is of course much less high; but probably it is often in a less cleanly condition than that of the country house.

Whatever may be the conditions of the building, a chimney shaft or flue, having its mouth and sides habitually open to the outer air, may be looked upon as forming as it were a small quadrangle or court within the enclosure of the building; and thus the sides of the shaft may be considered as *exterior* surfaces of the building, and as forming a local dielectric, consisting of a sheet of soot, to the plate formed by the ground at or near the foot of the shaft, which ground we must presume is more or less open through the shaft to the condensing influence of the clouds.

We have said that the grates are usually insulated from

[1] VI. D *a.*

VII. B *k*.

the ground by the hearthstones; and as the latter do not often appear to be recorded as being pierced when a chimney is the scene of a thunderbolt, probably the potential of the explosion accumulates on the ground immediately surrounding the front and back hearthstones, this accumulation being due to the loss of capacity caused by the facilitative dielectric formed by the tall column of soot,[1] aided as it frequently is by the grate, fire, and smoke.

The explosion, taking in its course any human body that may be near the line of least restraint, leaps from the ground near the hearthstones to the grate, and then ascends the chimney-shaft through the facilitative soot which lines it, and emerges at the pot or other summit of the stack.

The explosion frequently shatters this summit on leaving it, causing some of the débris to fall down the chimney-shaft.

The longer the shaft, the larger its girth, and the sootier its lining, the greater would appear to be the danger.

In nearly all private houses and cottages which do not contain much metal on their exterior surfaces, the chimneys form the probable *loci* of any explosions that may visit them, for not only is there danger from the soot-lined flues, but the stacks or pots crowning them usually constitute the most elevated features of the buildings.

Particularly at country buildings does injury from lightning occur at the chimneys; hence in devising measures for the defence of life and property from the action of lightning, the case of these buildings would appear to demand special attention.[2]

The chimney-stalks of furnaces would, from their great height, independently of their interior linings, be always apt to form dangerous local dielectrics.[3]

VII. B *l*; C.

(*l*) *Simultaneous Strokes of Lightning.*

When a building is struck simultaneously in several places apart from each other, the term "bifurcated," or "divided," is usually applied to the stroke.[1]

From what, however, has been advanced on the subject of explosions springing from the earth, there will be no difficulty in conceiving several such explosions occurring near each other simultaneously, at any place, and particularly at ordinary buildings, which frequently present more than one easy path for discharge around their walls.

These separate explosions seem especially likely to happen at buildings where the ground surrounding them is not in continuous electrical contact, but is broken up into insulated patches.

These strokes should thus apparently be characterized as simultaneous or multiple, rather than as divided.[2]

(C.) ANALYSIS OF THUNDERBOLT INCIDENTS.

With reference to the section just concluded, and in order to permit of a more ready comparison of the theories advanced therein with the facts related in Chapter III., an analysis will now be made of the incidents given in that chapter, arranged under the following heads, viz. :—

(1.) Buildings.
(2.) Ships.
(3.) Metals.
(4.) Chimneys.
(5.) Trees.
(6.) Flagstaffs, masts, &c.
(7.) Telegraphs.
(8.) Human beings.
(9.) Animals.
(10.) Simultaneous strokes.
(11.) Repeated strokes.
(12.) Accurately limited strokes.
(13.) Horizontally directed portions of strokes.
(14.) Acts of mechanical force.
(15.) Local plates.
(16.) Associated local plates.
(17.) Accidental dielectrics.
(18.) Local dielectrics.

---

[1] I. D 33, 34.   [2] VII. C 10.

LIGHTNING.

VII. C 1 a—ε.

Incidents at which lightning rods were present are specially considered in Chapter VIII.

(1.) BUILDINGS.

The following are the incidents referring to buildings:—

(a) *Town Churches.*

Nos. 36, 78, 107, 142,
51, 84, 118, 144,
52, 89, 132, 145,
53, 91, 133, 155,
59, 97, 134, 188.
62,

(β) *Town Public Buildings.*

Nos. 7, 58, 141, 178,
45, 127, 143, 191.
56, 136, 159,

(γ) *Town Private Buildings.*

Nos. 6, 18, 86, 137, 195,
8, 38, 92, 158, 196,
14, 40, 110, 189, 197.
17, 64, 126, 194,

(δ) *Country Churches.*

Nos. 11, 106, 138, 156,
12, 108, 139, 157,
13, 111, 140, 166,
16, 128, 146, 172,
24, 130, 147, 175,
41, 131, 150, 185,
68, 135, 153, 192.
79,

(ε) *Country Public Buildings.*

Nos. 39, 73, 114, 180,
46, 91, 121, 181.

VII. C 1 ζ—μ.

(ζ) *Country Private Buildings.*

Nos. 9,	35,	103,	177,
19,	37,	105,	183,
21,	49,	115,	184,
32,	55,	119,	186,
33,	66,	120,	200,
34,	67,	160,	201.

(η) *Labourers' Cottages.*
Nos. 167, 182, 203.

(θ) *Powder Magazines.*

| Nos. 27, | 47, | 84, | 129, |
| 28, | 82, | 87, | 179. |

(ι) *Barns, Sheds, and Outhouses.*

Nos. 26,	169,
113,	190,
163,	197.

(κ) *Tents.*
Nos. 88, 173.

(λ) *Miscellaneous Constructions.*

Lighthouse	.	.	.	.	No. 44
Windmill	.	.	.	.	42
Monument	.	.	.	.	74
Chimney Stalk	.	.	.	.	96
Gas-holder	.	.	.	.	159
Oil Tank	.	.	.	.	110[1]

(μ) *Buildings set on fire.*[2]

Nos. 53,	110,	141,	175,
83,	111,	154,	203.
84,	121,	163,	
87,	129,	172,	

---

[1] II. E 34.   [2] I. G 7, 17, 21.

212 LIGHTNING.

VII. C 1 ν, π; 2 a.

(ν) *Injuries inside Buildings.*   (π) *Masonry rent.*

Nos. 3,	51,	158,	Nos. 11,	97,	155,
4,	52,	159,	12,	106,	157,
13,	62,	167,	13,	107,	160,
14,	67,	177,	16,	108,	166,
18,	83,	178,	19,	111,	172,
19,	84,	181,	20,	112,	177,
23,	87,	182,	21,	114,	180,
26,	88,	183,	24,	118,	181,
27,	111,	184,	39,	119,	182,
28,	112,	186,	43,	126,	183,
32,	113,	189,	51,	128,	184,
33,	115,	190,	52,	130,	186,
34,	118,	191,	55,	131,	189,
35,	119,	194,	62,	132,	192,
37,	126,	200,	68,	133,	196.
38,	129,	201,	73,	134,	
40,	134,	203.	91,	137,	
45,	157,		96,	147,	

The cases of buildings struck mentioned in the above incidents are remarkable chiefly from the lightning rods, metals, or other important circumstances that are recorded to have been present at the time. These circumstances are referred to more fully under their respective heads, the question of lightning rods generally being dealt with in Chapter VIII.

As regards the small proportion of incidents in which buildings of brick and stone are related to have been struck, but no metals are mentioned, there is very little doubt, from the nature of these buildings, that they all contained more or less external metal.

(2.) SHIPS.[1]

(a) *In motion.*

Nos. 5,	75,	99,	104,
63,	76,	100,	125.
70,			

[1] I. G 1—5, 37.

THE ACTION OF THUNDERBOLTS.   213

VII. C 2 β; 3 a—δ.

(β) *In harbour.*

Nos. 48,	61,	71,	101,
50,	65,	72,	102,
60,	69,	92,	109.

The more remarkable of the above incidents are those at which lightning rods were present; and these are treated of in Chapter VIII.

(3.) METALS.

(a) *External Copper, Bronze, Brass, or Gilding.*

Nos. 7,	62,	103,*	131,*
16,	78,*	107,	132,
36,	79,*	111,	133,
51,	81,	117,	142,
52,	96,	128,*	157.*
61,	97,		

(β) *External Iron.*

Nos. 15,	63,	129,*	145,
37,	68,	131,*	154,
38,	81,	132,	155,*
41,	86,*	135,	178,
42,	106,	136,	180,
47,	112,	138,	183,
51,	114,	139,	188,
52,	115,*	140,	191,
60,	117,	143,	195,
61,	119,	144,	198.
62,			

(γ) *External Lead.*

Nos. 47,	58,*	97,	133,
51,	62,	131,*	180.
52,	86,*	132,	

(δ) *Internal Copper, Bronze, Brass, or Gilding.*

| Nos. 9, | 17, | 88, | 191, |
| 13, | 18, | 178, | 203. |

VII. C 3 ϵ.

(ϵ) *Internal Iron.*

Nos. 17,	67,	118,*	159,	191,
19,*	83,	157,*	178,	196,
20,	88,	158,*	181,	203.
45,*				

In cases marked thus * lightning rods were also present.

We have seventy-three separate cases here recorded in which metal exerted its attractive or accumulative qualities to bring about explosion. These cases are well worth studying, though the details of them are unfortunately, in some instances, rather scanty.

Those which deserve special attention, either from their remarkable character or on account of the fulness of the details afforded, are as follows:—

No. 19. Mr. Raven's house and fowling-piece.
20. The soldiers at Martinique.
41. Lausanne Cathedral.
45. Charlestown Prison.
51. St. Martin's Church.
52. Brixton Church.
62. St. Bride's Church.
88. Sutton Camp Mess Tent.
96. Royal William Victualling Yard Chimney Shaft.
106. Black Rock Church.
112. House at Concordia.
117. Paignton Flagstaff.
129. Bruntcliffe Powder Store.
132. St. George's Church, Leicester.
133. Merton College Chapel.
157. Clevedon Church.
158. The banker's house at Lyons.
159. Halifax Buildings.
178. Telegraph School at Malta.
203. Cottages at Ide.

The chimney-shaft case, No. 96, deserves the more notice, since it is one of the instances that have been put forward by authorities as showing that lightning is not necessarily

VII. C 4, 5.

attracted by metal, on the ground that there was none in the shaft itself, whilst a tower some distance off contained a great deal and was not struck. The influence of the large copper roof just below the shaft, where the explosion is recorded to have ended (but where it probably in reality began), is not, however, mentioned by these authorities.

In connection with these metallic incidents, two not included in the foregoing summaries, viz. Nos. 94 and 95, are noticeable, as recording how in their cases metal buildings had for centuries not been struck.

In one case, however, No. 94, the presence of numerous metal points is mentioned, and there is no doubt that the other building also contained many angularities and probably some well-defined points; whilst both appear to have had good metallic connection with the ground; so that the element of leakage[1] doubtless, in both instances, overcame that of accumulation, and the result was that, *quâ* their metal, these buildings were not found to be sources of danger.

(4.) Chimneys.

Nos. 55, 73,	119, 126.	167, 177,	181, 189,	196, 200,	203.

The following are noticeable:—

> No. 167. Lossiemouth Cottage.
> „ 177. Mr. D. Onslow's house.
> „ 203. The cottage at Ide.

(5.) Trees.

Nos. 2, (e) 10, (o) 25,	29, (o) 49 93, (o)	122, 124, 149,	152, (p) 162, (e) 164,	165, 173, 174, (c)	187, (e) 193, 198.

N.B.—(e) denotes elm; (o) oak; (p) poplar; (c) cherry.

---

[1] VII. B *d*; F *e*.

216 LIGHTNING.

VII. C 6—8 a—γ.

(6.) Flagstaffs, Masts, Roofs, and Wooden Objects generally (exchpt Ships' Hulls).

Nos. 10,	47,	103,	129,	175,
15,	51,	104,	141,	180,
16,	60,	109,	146,	185,
25,	61,	111,	157,	188,
27,	63,	112,	161,	194,
28,	71,	117,	169,	201,
32,	83,	120,	170,	203.
42,	88,	125,	172,	

The injuries to flagstaffs would seem to be lessons pointing out their dangers to those who are in the habit of using them as ornamental appendages to their residences,[1] or as architectural features of church towers.[2]

(7.) Telegraphs.
Nos. 178, 191.

(8.) Persons.[3]

These are all fatal cases, except those marked thus *, which are instances of shocks. Those marked thus ‡ are cases of fatalities as well as shocks.

(a) *Inside houses in towns.*[4]

Nos. 26,	35,	45,*
32,	38,	134.*
33,	40,	

(β) *In the open air in towns.*
Nos. 23, 148,* 168.‡

(γ) *Inside houses, and under cover, in the country.*[5]

Nos. 3,	88,*	169,	182,
4,	113,*	173,*	190,‡
34,	115,‡	177,	200,*
35,	167,	181,*	203.*
37,*			

[1] 117. [2] 185. [3] I. G 6, 21.
[4] I. G 13. [5] I. G 13.

THE ACTION OF THUNDERBOLTS. 217

VII. C 8, δ—o.

(δ) *In the open air, in the country.*[1]

Nos. 1,	90,	129,	164,
10,	98,	149,‡	165,‡
15,	116,	151,	170,
20,	117,*	161,	171,‡
25,	122,	162,‡	193.
82,*			

(ε) *In ships.*[2]
No. 72.*

(ζ) *Under water.*
No. 123.*

(η) *Riding or driving.*
Nos. 15, 161.

(θ) *Walking.*
Nos. 116, 168,‡ 170.

(ι) *Engaged in agricultural pursuits.*
Nos. 90, 165,‡ 170, 171,‡ 190.‡

(κ) *On railways.*
Nos. 98, 151.

(λ) *Sheltering under trees.*[3]

Nos. 10,	149,‡	165,‡
25,	162,‡	193.
122,	164,	

(μ) *In barns, outhouses, or tents.*
Nos. 26, 88,* 113,* 169, 173,* 181,* 190.‡

(ν) *In labourers' cottages.*
Nos. 167, 182, 203.*

(o) *In bed.*[4]
Nos. 32, 33, 34, 35, 203.*

[1] VII. B ε.   [2] I. G 13.   [3] I. G 13.   [4] I. G 13.

L

VII. C 8 π—τ, 9, 10.

(π) *In church.*[1]
Nos. 3, 4, 134.*

(ρ) *Experimenting with lightning rods.*
No. 38.

(σ) *Soldiers on duty.*
Nos. 20, 82.*

(τ) *Clothes of persons.*
Nos. 26, 161, 162, 165.

The following are the cases worthy of the greater attention, viz. :—

    No. 15. The Coldstream carter.
        38. Professor Richmann.
      115. Mr. Buys.
      161. Mr. Woodman.
      167. Mrs. Whyte.
      168. The persons in Victoria Park.
      171. The lads in Beresford's Fields.
      195. The platelayer on the Midland Railway.
      203. Mr. H. S. Stobart.

From No. 168 we see how an open park in the midst of an immense city is chosen as the scene of a fatal explosion, rather than the city itself; and this tends to show that residents in the country are, generally speaking, more exposed to death by thunderbolts than the denizens of towns.[2]

(9.) ANIMALS.
Nos. 15, 30, 31, 85, 113, 116, 161, 176, 182, 187, 199, 202.

(10.) SIMULTANEOUS STROKES.

Nos. 47, 103, 108, 128,
72, 104, 119, 156,
73, 105, 120, 203.
91,

[1] I. G 13.   [2] I. G 11, 35, 36.

(11.) Repeated Strokes.

Nos. 5,	57,	77,	81,
13,	63,	78,	121,
14,	72,	79,	155,
53,	76,	80,	156.

(12.) Accurately defined Strokes.

Nos. 2,	33,	67,	171,
27,	34,	93,	182.
28,	37,	114,	
32,	54,	167,	

(13.) Horizontally directed Portions of Strokes.

Nos. 47,	117,	157,	182,
54,	156,	178,	203.

[N.B.—The five following groups do not include cases of lightning rods.]

(14.) Acts of Mechanical Force, exclusive of bending of Masonry.

Nos. 2,	33,	103,	156,
5,	51,	109,	166,
11,	52,	112,	173,
12,	54,	117,	174,
15,	60,	119,	180,
16,	61,	120,	181,
24,	62,	124,	182,
25,	63,	125,	183,
26,	67,	128,	185,
27,	71,	138,	187,
28,	75,	139,	195,
29,	86,	146,	196,
32,	88,	152,	197.

(15.) Objects which, when struck, probably formed Local Plates.

Nos. 1,	40,	98,	142,	162,	176,
3,	50,	110,	143,	164,	190,
4,	63,	116,	144,	165,	193,
10,	76,	122,	145,	168,	198,
23,	82,	135,	149,	169,	199,
30,	85,	136,	151,	170,	202.
31,	90,	140,	154,	171,	

VII. C 16—18; D.

(16.) Objects which, when struck, probably formed Local Plates associated with Local Dielectrics.

Nos. 20,	52,	97,	120,	138,	158,
39,	88,	114,	132,	139,	178,
51,	96,	117,	134,	147,	191.

(17.) Objects which, when struck, probably formed Local Plates, accidentally constituting Local Dielectrics.

Nos. 15,	48,	61,	75,	109,	125,	152,	173,
25,	56,	63,	93,	113,	127,	159,	174,
26,	60,	71,	104,	124,	129,	161,	187.
29,							

(18.) Objects which, when struck, probably formed Local Dielectrics.

Nos. 2,	27,	54,	107,	141,	177,	194,
7,	28,	67,	111,	146,	180,	195,
9,	32,	68,	112,	155,	181,	196,
11,	33,	73,	115,	156,	182,	197,
12,	34,	74,	118,	157,	183,	200,
13,	35,	81,	119,	160,	184,	201,
14,	36,	83,	121,	163,	185,	203.
16,	38,	84,	126,	166,	186,	
17,	41,	87,	130,	167,	188,	
18,	42,	105,	133,	172,	189,	
24,	53,	106,	137,	175,	192,	

(D.) Atmospheric Dielectrical Conditions.

We propose now to treat of the principal special conditions of the air dielectric that would appear to affect explosion,[1] so far as these conditions may be said to be independent of the question of electrical leakage.

This question, which would be of great importance in the consideration of the state of the air previous to the development of thunderstorm elements, is dealt with in Section F of this chapter.

---

[1] III. 1—4, 15, 50, 85, 94, 117, 157, 178, 183.

VII. D a.

(a) *The Influence of Rainfall.*

A condition almost always present with thunderbolt explosions is rainfall, and its action deserves, therefore, special attention.

The remarks applied hereafter to rain would also apply, *mutatis mutandis*, to the occasional substitutes for rain, viz. snow and hail.

The under-surfaces of the clouds, whence rain springs, being rendered more moist by its eruption than they were before, must become more collective, and, therefore, on the principle submitted in Section A, must exert in a given time a proportionally greater condensing power on the potential of the earth's collecting plate.[1]

The air dielectric being impregnated with the rain drops becomes more facilitative of explosion, *i.e.* exercises less restraint on the recombination of the electricities accumulated on the two plates; or, to express it differently, capacity being reduced and quantity unaffected, potential is increased.[2]

Lastly, the earth's surface becomes more collective owing to the access of moisture it receives, and therefore its potential, like that of the clouds, is increased.[3]

Thus we see that in three distinct ways rainfall tends to increase potential and to bring about explosion, and, taking all the circumstances into consideration, we may probably conclude that, as a rule, rainfall is the immediate practical cause of the occurrence of thunderbolts.

In accordance with this idea the following three conditions would probably be those which would generally precede these explosions, viz.: First, a collective portion of the earth's surface already charged from below with a certain quantity of electricity. Secondly, a low dense rain cloud approaching the zenith of the charged place, and

---

[1] VI. D *a*.   VII. A *a*.   [2] V. B 1, 7, 12.
[3] VI. D *a*.   VII. A *a*.

VII. D b.

ready on arrival within inductive range of it, to increase its potential by condensation. Thirdly, a fall of rain suddenly reducing the capacity of the air whilst the accumulation of potential on the cloud and the place is proceeding; this rainfall being, in fact, the last straw that breaks the camel's back.[1]

The density of the rainfall would doubtless have an effect on the restraint of the air;[2] and a thunderstorm having once commenced, it seems reasonable to expect that the heavier the rainfall that should accompany it, the more numerous, *cæteris paribus*, would be the thunderbolt explosions that would occur during its continuance.

(b) *The Temperature of the Air.*

The restraining power of the air is probably influenced in some degree by its temperature; but on this, as on the question of the influence of the density of rainfall, there appears to be but little knowledge extant.

Thunderbolts occur more numerously in this country during the warm than during the cold season; but they are by no means absent during the latter.[3]

They occur frequently in all countries and regions at night-time;[4] though in this country their appearance is generally by day.

The reason for the formation of snow and hail is of course associated with atmospheric temperature; and when hail accompanies thunderstorms, they are said to be usually of a severe character;[5] but the connection between hail and thunder storms has not been well accounted for.

In England we probably suffer far less from hail than is the case in France, where, in the vine regions, it appears

---

[1] I. D 18. II. G 29.   [2] VI. D a.
[3] III. 12, 24, 44, 46—48, 59—61, 72, 76, 106, 117, 127, 135, 144, 147, 182.
[4] III. 32—35, 53, 106, 112, 125, 181.   [5] III. 117. I. C 23.

VII. D c; E a.

to constitute a dreadful scourge, and to be generally accompanied with much thunder and lightning.[1]

It was thought by Arago that the same means that would tend to prevent thunderbolts might also tend to prevent hail;[2] and it certainly appears possible that increased researches as to the action of lightning may also throw light on that of hail.

(c) Atmospheric Electricity.

The presence of charge in any portion of the dielectric, independently of that on the plates, would probably tend to affect its restraining power.[3]

It appears from experiments that have been made that the stratum of air lying nearest the earth is generally more or less negatively electrified.[4]

Thus the lower regions of the atmosphere may possibly, from this cause, possess in some places degrees of restraint different from what they do in others.

It can be conceived that the presence of this atmospheric electricity in any large amount on or over any particular portion of the terrestrial plate might materially facilitate explosion thereat, but our knowledge of atmospheric electricity appears to be so slight that it is impossible to come to any conclusions on the subject.[5]

(E.) CLOUDS AND CLOUD EXPLOSIONS.

We will now notice the principal conditions, tending to charge and discharge, connected with the clouds; and here we enter into a consideration of the discharges that occur between clouds and clouds, viz. cloud explosions.

(a) The Electrical Conditions of the Clouds.

The charging to some extent of the surface of the ground from the action of electricity below it (due to causes

[1] I. C 21. [2] I. C 22. [3] I. C 10—14.
[4] I. C 8, 9. [5] I. C 2, 3.

VII. E b.

of which we are as yet ignorant) being the first step towards the formation of a thunderstorm condenser, the second step is probably the piling up of clouds, at no great elevation, over a particular area.[1]

The causes of the collection of clouds in masses suitable to the development of these condensers are probably of a purely meteorological nature.

A condenser having been formed, the condensive power of the condensing plate formed by the under-surfaces of the clouds would be an element affecting explosion.[2]

This condensive or collective power would probably be influenced by the amount of moisture[3] present, and by the cloud's area and shape.

The effect of rainfall in increasing the moisture of the under-surfaces of the clouds has already been suggested; and, evidently, the greater the area of this surface and the more parallel it lies to the earth's surface, the greater is the chance of any charge already collected on the latter being condensed by the cloud during its passage over it.

(b) *The Electricity due to the Conversion of the Clouds into Rain.*

It seems probable that the potential of the under-surfaces of the clouds is affected, independently of their induction by the earth, by a disengagement of electricity due to the sudden conversion of the lower portions of these dense masses of vapour into rain.[4]

The electricity thus formed would doubtless go to swell the induced charge already existing; for if it were of an opposite nature the two electricities would tend to neutralise each other, and thus rainfall would tend to prevent thunderstorms, which is a very unlikely state of things.

---

[1] I. C 27—30, 36; D 27.
[2] VII. A a.
[3] VI. D a.
[4] I. C 16, 17, 118.

VII. E *c*.

(*c*) *Cloud Explosions.*

A cloud explosion is the explosion of a condenser formed by two separate clouds, one of which, constituting the collecting plate, has received its charge by induction from the earth.[1]

The degree in which the clouds are broken up into detached masses must be a matter of great moment; for, if they constitute a single huge plate, there is no possibility of explosion taking place from it except with the earth.

But if there are a number of separate insulated clouds, a great chance then exists of these clouds becoming discharged by explosions with each other instead of with the earth, or, in other words, of cloud explosions occurring in lieu of thunderbolts.[2]

The main factor in the problem whether, in the case of a number of detached clouds, explosions would occur between them or with the earth, would be their relative distances at any moment from each other and from the earth.

The mobility of these light masses of vapour renders it likely that they would frequently sail within explosive range of each other before the necessary conditions for their explosion with the earth had ripened.

The directions, then, in which they might move would be very material circumstances, and the great causes influencing these directions would appear to be three in number, viz. :—

    (1.) Wind, or air currents.
    (2.) Electric attraction and repulsion.[3]
    (3.) Gravity.[4]

Our knowledge of the nature of air currents would lead to the belief that their force would generally act rather in

---

[1] I. D 10, 16. VI. E *a*.   [2] I. C 33. VI. A *a*.
[3] I. C 31, 32. V. A 7, 8.   [4] I. C 31, 32.

VII. E d.

a lateral than in a downward direction, and therefore would tend to promote cloud explosions rather than thunderbolts.

Electric attraction would, *quâ* the earth's influence, be a downward force; whilst both attraction and repulsion from the adjacent clouds would act more or less laterally.

Thus when the clouds were consolidated in one great plate, the earth's attraction would be a considerable element in hastening thunderbolt explosions; but when there were several clouds floating about, it would be difficult to conjecture what the net result of all the electric attractions and repulsions would be in hastening or delaying such explosions.

Gravity of course acts in a direction tending to cause thunderbolt explosions.

It would appear, then, that during a thunderstorm the circumstances of the clouds are considerable elements in determining whether the explosions that mark its progress will take the form of cloud explosions or of thunderbolts.

(d) *Thunderstorms.*

The subject of the motion of the clouds leads to a consideration of the motion and cause of thunderstorms.[1]

Thunderstorms, having commenced, generally move in a certain direction, over a certain area or region, before they disappear.[2]

The explanation of the commencement and progress of a thunderstorm is probably as follows:—The clouds which cause the storm move, more or less together, in a particular direction (mainly that of the wind), and this direction happens to be over ground that is previously charged from below.

The storm thus becomes a thunderstorm; and the clouds during their progress are repeatedly electrified by induction by the fresh places, most favourably adapted for this

---

[1] I. G 38.   [2] I. C 43, 44.

VII. F a.

action, whose zeniths they are constantly approaching; and each electrification results sooner or later in a cloud explosion or a thunderbolt.

As soon as these clouds arrive over uncharged ground, or mount higher in the atmosphere, or disperse (the rain ceasing in both these latter cases), the thunderstorm of which they are the immediate cause, ceases, and the atmosphere resumes its normal state.

There is apparently no reason for assuming that the surface of the earth is not frequently charged from below at times when no thunderstorms happen, and when the weather is perfectly fine;[1] and these storms are probably due to the fortuitous conjunction of low piled-up masses of rain cloud and charged surfaces of ground.

(F.) TERRESTRIAL RETURN STROKES.

(*a*) *Nature of Terrestrial Return Strokes.*

Terrestrial return strokes, or induced discharges of terrestrial electricity, affect the surface of the earth generally, and human beings, animals, and telegraphs in particular.[2]

The conditions necessary to induce return strokes are of two kinds, viz. :—

> 1st. The charge on any portion of the earth's surface, or collecting plate, being condensed by a cloud, and an explosion occurring between this cloud and another one; in other words, whenever a cloud explosion occurs.
>
> 2nd. The charges on two or more portions of the earth's surface, or collecting plate, adjacent to, but insulated from, each other, becoming condensed to different degrees of potential by the same cloud, and an explosion happening between

[1] I. C 3.
[2] I. D 51—58, 61—63, 65—71, 76. V. B *e*. VI. E *a*.

VII. F b.

this cloud and that portion of ground whose charge was at the highest potential; and this would, in fact, generally be whenever a thunderbolt occurred.

The portion of the earth's surface in the first case, and the unexploded portions in the second, must then necessarily lose so much of their potential, and therefore (capacity being unaltered) of their charge, as was due to the condensing action of the cloud before, in each case, it exploded.

These charges thus lost return to meet and recombine with the oppositely electrified particles which they had temporarily displaced and driven into the earth when they were drawn to and collected on the surface by the cloud's condensing action; and thus terrestrial return strokes are occasioned.

(b) *Return Strokes induced by Cloud Explosions.*

Return strokes of the first order, viz. those induced by cloud explosions, must necessarily be beneficial to the earth's surface and to its inhabitants; for the oppressive feeling that occurs shortly previous to, and during the prevalence of, thunderstorms—a feeling due to the accumulation of potential brought about by the condensing action of the clouds—is removed, the air is (to use the popular expression) "cleared," and the slow chemical decomposition which manifests itself during this period in certain substances, such as milk and beer, a process likewise due to the abnormal amount of potential present, ceases.[1]

Whenever, then, cloud explosions happen, *i.e.* when lightning appears in the heavens only, and not at the earth, corresponding return strokes take place on the earth's surface, such strokes being chiefly manifested by their general *malaise*-removing results.

[1] I. C 19, 37, 38.

VII. F c.

(c) *Return Strokes induced by Thunderbolts.*

Return strokes of the second order, though induced by thunderbolts, are never likely to be so dangerous as these explosions; for there is no scope in the nature of return strokes for violent manifestations, since they can only occur within a collector (the presence of a non-collector being, as we have seen, essential to explosive action), and, in leaving that collector, they return *through* it, instead of going out of it into a non-collector, as a thunderbolt does.

The effect of a thunderbolt-induced return stroke on a human being is usually a shock to the system,[1] the effects being generally temporary, and lasting not more than a few hours; and there appears to be little, if any, ground for deeming that this shock is ever of a fatal nature; for when a person is struck dead by lightning, it must be generally quite out of the power of any bystanders, or witnesses (who cannot possibly have been expecting the stroke), to give an opinion as to how the event occurred; and any such evidence as to the fact of fatal return strokes can hardly be conclusive.

As regards the Coldstream incident,[2] which has been attributed to a return stroke,[3] the fact of thunder accompanying it would alone appear sufficient to render such an idea nugatory; whilst the circumstance that two persons who happened not to be far off at the time stated that they had seen no lightning, does not preclude the strong probability that there was lightning of some kind, and during some portion of the thunderbolt's passage through the air, especially when we remember that lightning necessarily requires a certain condition of atmosphere for its manifestation, and that it is possible that in certain weathers this state is not always present in every stratum of air through which an explosion may have to pass; thus the phenomena of "fireballs" are probably manifestations

VII. F d.
of thunderbolt lightning under abnormal atmospheric conditions.[1]

Return strokes are generally experienced, as might be expected, by persons in the immediate vicinity of the scene of a thunderbolt explosion; and in these cases it is often remarkable how the thunderbolt selects one person as its victim, whilst others, who may be close to this person at the time, are either totally unaffected or merely receive shocks.

This shows within how minute an area a thunderbolt occurs, how exactly it chooses its place and path, and how circumstances that are generally regarded as trivial (for no attempt seems ever to be made in this country to investigate the reasons for such selections) doubtless make all the difference whether a person is killed or not.[2]

(d) *The Effect of Return Strokes on Telegraphs.*

When telegraph wires form part of the collecting plate whose condensed electricity departs in a return stroke, whether such be induced by a cloud explosion or by a thunderbolt, these wires are apt to be traversed by currents of more or less strength, which are liable to injure the instruments at the stations.[3]

Apparatus called "lightning protectors" have, however, been devised, which prevent much practical inconvenience to telegraphy from the effect of these abnormal currents,[4] though they are less to be depended on for protecting the instruments from thunderbolts.[5]

When the cables of submarine mines, or the wires of land mines and blasts, become the *loci* of return strokes, the induced currents set up are clearly sources of the utmost danger, and need to be guarded against with the greatest vigilance.[6]

[1] I. D 4, 25, 26.   III. 50, 85, 178, 183.   VII. C 12.
[3] I. D 66—70.   [4] I. D 74, 75, 77.   [5] VII. C 7.
[6] I. D 71, 76.

VII. G a.

The question of the interference of thunderstorm electricity with artificial electrical arrangements is obviously one that forms part of the practical science of these arrangements; thus any detailed consideration of it hardly comes within the limits of this treatise.

(G.) TERRESTRIAL LEAKS.

We have already submitted that electric leaks are of two kinds, the one due to the porosity of all insulators, and the other to the existence of angularities on the surface of collectors.[1]

Leakage of either kind from particular places on the earth's surface,[2] through the air, prior to the formation of the cloud condensing plate, would evidently materially affect the collection of electricity at such places, and would probably sometimes determine whether a storm passing over them became a thunderstorm or not.

Terrestrial leakage is thus an element of great importance, and its presence must tend considerably to lessen the chances of explosion.

We will first treat of the principal sources of porosity in the atmospheric insulator, and then of the angularities that occur on the terrestrial collector.

(a) *Atmospheric Porous Leaks.*

Smoke and foreign gases in the layer of atmosphere immediately overlying the earth appear to have the property of more or less piercing this layer.

Hence, large cities where coal is burnt, the smoke of which usually forms a canopy over them, are probably by this circumstance defended to some extent from thunderbolts.

The density of the atmosphere would seem to be another agent affecting leakage, since the pores of the air may be considered open in inverse proportion to its pressure.

VII. G b.

The state of moisture of the atmosphere has probably considerable influence on leakage; and a saturated condition of the air would doubtless tend to increase its porosity.

Aurora, and heat (or summer) lightning would appear to be luminous manifestations of porous leaks in the atmosphere; the one, between the earth and the upper atmospheric strata,[1] and the other, between two of these strata.[2]

Waterspouts appear to be mechanical manifestations of atmospheric leaks; for under circumstances which show that electricity is present, we see clouds depressed, and the movable surface of the earth, *i.e.* waters, elevated to meet them.[3]

(b) *Terrestrial Angular Leaks.*

When we come to those sources of leakage which are derived from conditions appertaining to the earth's surface, we are on much surer ground than we can ever be in the case of the atmosphere.

There is no doubt as to the fact that projecting angularities on the surfaces of collectors lose their capability of holding charge in proportion to the collectivity of their material and to the acuteness of their projections; whence it follows that a sharp metallic point on a collector constitutes a great leak, and the discharge it occasions is of the nature of a continuous escape of electricity from the collector, through a rift in the enveloping insulator, to meet the induced electricity which necessarily flows towards it from the nearest separate collector.[4]

Practically, the only angularities that materially exercise this function are those made of metal, other substances not having sufficient accumulative power, *i.e.* the particles of electricity gathered on them are not able to exert a suffi-

---

[1] I. C 78. VI. A *a, b ;* E *a.*   [2] I. D 4, 20. VI. A *a ;* E *a.*
[3] I. C *i.*   [4] V. B 18.

VII. G *c*.

ciently powerful repulsive or dispersive action on each other.[1]

Now, as the earth's surface does not present naturally any metallic angularities, although such projections as are offered by blades of grass, foliage, thorns, branches of trees, and shrubs, and even sharp-pointed rocks, do doubtless act in some slight degree as leaks, we are, in order to obtain dispersers of practical value, compelled to use artificial means.

By the employment, then, of such artificial means as metal points, we have the power of tapping the electricity of the earth, and of assisting to rid its surface of the charge that may collect thereon from below.

(*c*) *The Value of Metal Points in Relation to the Earth.*

If we assent to the theory that the clouds form the collecting plate, and that the earth's surface receives its electricity therefrom by induction,[2] we must relinquish the idea of dispersing the terrestrial electricity until the thunderclouds have actually arrived and the process of condensation has commenced.

Now, although, at that time, metal points on the earth's surface would undoubtedly have a considerable duty to fulfil in ejecting the earth's charge as quickly as they could, and in helping to dissipate that of the thunderclouds, yet this duty would be far less valuable than that which these points would carry out if the earth's surface should form the collecting plate; for in that case they would cause leaks from the original charge, and would reduce its quantity and potential *before* the thunderclouds arrived and the condenser became formed.

In the case first supposed there is reason to believe that, after the clouds had begun their condensing operation, potential would be liable to accumulate with such rapidity

VII. G c.

that, however effective the points might really be, they would not be able to throw off the charge at an equal rate, and thus explosion would not in most cases be prevented;[1] but if the earth form the originating plate, and if the ground have been thoroughly tapped *beforehand*, there would be little or no charge left on it to be condensed when the clouds arrive, and consequently explosion would be much less likely to occur.[2]

It is thus evident that when the earth is treated from the point of view upheld in this treatise, viz. as the collecting plate of the terrestrial condenser, and the originator of thunderstorm electricity,[3] the value of metal points in tending to prevent thunderbolts becomes greatly enhanced.

We look on the earth, then, under this aspect, as a receptacle of force in a more or less pent-up state, and, like all such receptacles, as needing valves or outlets for the escape of such force before it becomes explosive.[4]

The medium by which, in the case of the earth, this explosive stage is liable to be reached, is the thundercloud; but this, fortunately, only comparatively seldom exists; and, between its intervals of existence, it is necessary to tap the earth's surface, or those portions of it which we are particularly interested in protecting, in order that when this cloud *does* appear it may have little or nothing to work upon; and thus the terrible result of its work, the thunderbolt, may be averted.

The luminous phenomena known as St. Elmo's Fires, which, in thunderstorm weather, appear at the upper extremities of pointed metals connected with the earth, are tangible proofs of the tapping powers of metal points;[5] so also are the brush discharges artificially produced from points on charged metallic collectors;[6] and the flashes that were drawn by the pioneers of lightning engineering

---

[1] II. C 45. [2] II. C 11, 30, 51, 53—55; G 40. [3] VI. A *b*.
[4] II. E 27. [5] I. C *h*. [6] II. C 11, 31, 52, 53.

VII. G *d.*

with their pointed rods give testimony to the same effect.[1]

(*d*) *Terrestrial Valves afforded by Features of Civilisation.*

We propose to notice now some of the principal valves to terrestrial electricity afforded by various features of civilisation, apart from any apparatus specially devised to defend life and property from the action of lightning; and our attention will first be directed to the case of large towns.

We have already alluded to some probable reasons for the apparent comparative immunity possessed by large towns in regard to thunderbolt injuries, viz. the absence of open fields and vegetation, the pavements immediately surrounding the houses, and the smoky atmospheres.[2]

Another reason, however, and probably a still more efficient one, lies in the fact that the surface crust of the ground in all large towns is permeated by much metal in the shape of water-pipes and gas-pipes, and these pipes have all of them terminations, more or less angular, above ground, *e.g.* water-cocks and gas-burners within the houses, and the burners of gas standards in the streets.

We thus obtain two efficient systems of electrical taps, exposing comparatively little metallic surface to the action of the clouds, and preventing any charge that may rise to the earth's surface from accumulating at any particular place, by quickly dispersing such charge through the numerous angular exits afforded; for the ejective power inherent in such of these exits as occur *inside* buildings is not affected thereby.[3]

Towns furnished with water or gas supply would, therefore, seem to be well tapped, and to be fairly protected, both as regards their houses and the open streets, from injury by thunderbolts.

In the case of the country, the network of metal formed

---

[1] II. C 13, 14, 22, 23.   [2] VII. A *c, e;* G *a.*   [3] V. B 20.

VII. G *e*.

by the railway system probably constitutes an efficient source of leakage.[1]

The metals are practically connected with the earth's surface, and they afford, together with the engines running on them, numerous angularities through which the terrestrial electricity may escape.

The telegraph posts, which are now erected along the main roads, must also, by means of their earth wires, constitute good taps.[2]

It thus appears that four distinct features of civilisation, otherwise beneficial, viz. water supply, gas supply, railways, and telegraphs, tend also to protect life and property from thunderbolts.

(e) The Angular Leakage of Metals.

Allusion has already been made to the conditions of leakage inherent in all metal local plates.[3]

This is due to the impossibility of manufacturing metal articles without angularities, more or less sharp, in the shape of corners and edges.

Under this aspect, the metal objects that would furnish sources of leakage in a minimum degree would be flat roofs,[4] domes, balls, and curved surfaces.

Ordinary houses are probably occasionally protected to some extent by the fact of their rain-water pipes entering the ground at one or more places close to the outsides of their walls, these pipes either terminating above in hopper heads, or being connected to the eaves gutters, both of which furnish ridges and angularities.[5]

It has to be borne in mind that it is only when metals are metallically connected to the ground, and so form local plates, that points and angularities can be of any advantage to them; for there is no reason to suppose that, when

---

[1] VII. A *f*.   [2] II. B 28.   [3] VII. B *d*.
[4] III. 51, 52, 86, 96, 97, 115, 131—133. II. B 54.
[5] II. E 3, 8, 33.

VII. G *e.*

they form local dielectrics, any particular shape they may have is of the slightest consequence.

It will be evident, from what we have advanced, that all metals in local plates present two antagonistic elements, the one tending to accumulate charge, and the other to disperse it; and in the difficulty, and probably the impossibility, of deciding whether the dispersive qualities of a metal surface act at the same rate as its accumulative powers, lies the great danger of using metal at all on the outside of buildings, however well such metal may be connected to the earth.[1]

[1] I. E 1—12. II. G 3, 4, 8, 14, 15, 25—27.

VIII. A a.

CHAPTER VIII.—THE PRESENT SYSTEM OF LIGHTNING RODS.

(A.) OBSERVATIONS ON THE HISTORY OF LIGHTNING RODS.

(a) *The Invention of Lightning Rods.*

WE have now arrived at the stage for considering the means generally adopted for defending constructions, *e.g.* buildings, ships, and monuments, from the action of thunderbolts.

It has been shown how, through the ejective power of metallic points, we possess to some extent a grasp over the action of terrestrial electricity.

This power was discovered by Benjamin Franklin; and it enabled that great man to devise for the protection of buildings and other constructions the arrangement of pointed metal known as the lightning rod.[1]

A lightning rod, as invented by Franklin, and as now in general use,[2] may be said to consist of three parts, viz. :—

> (1.) A root, earth connection, or underground portion, which is usually adjacent to the base of the most elevated feature of the building.
> (2.) A stalk, *i.e.* a rod, band, tube, or wire rope, proceeding from the root more or less vertically up the outside of the building, and projecting some distance above its uppermost feature.
> (3.) A point forming the upper extremity of this stalk.

The fact that since Franklin's time no material alteration in the design or the construction of the lightning rod has

---

[1] II. A 12—16; C 14, 42.   [2] IV. 1—48.

VIII. A b.

been adopted, is a remarkable testimony to his genius; and the circumstance is all the more notable when we remember that the science of electricity was then in its infancy, and had not as yet been applied to practical purposes.

The lightning rod was probably the earliest application of electricity to a useful object.

The scope of lightning rods, however, although of a most beneficial nature, has never been connected with economical or commercial necessities, and therefore the art of lightning engineering has perhaps not received that full and searching consideration to which more business-like arts are generally subject.

The development of the electric telegraph, and of all the valuable employments of current electricity that followed in the wake of Volta's grand discovery, has naturally so absorbed the energies and researches of eminent physicists that a contrivance not intended to meet the wants of everyday life, but merely to provide an insurance (and, in the opinion of many, a very uncertain one) against a somewhat unlikely accident, has necessarily been somewhat eclipsed by more cogent interests.

It is submitted, however, that our present knowledge of the laws of electricity and of the terrestrial constitution, taken in conjunction with the mass of experience that has been furnished by the accidents occasioned by thunderbolts during the one hundred and thirty years that have elapsed since the introduction of lightning rods—and especially with that portion of this experience that bears on the influence of these rods—enables us now to rear a superstructure on Franklin's foundation, and to deal far more thoroughly with the subject than he possibly could have done.

(b) *Theory of the Functions of Rods.*

There seems but little doubt that Franklin's fundamental object was to elicit the charge from the ground by means

VIII. A b.

of the lightning rod's point,[1] and that the idea of offering the lightning an easy conduct to the ground by means of the stalk, was the result of a later conception of the theory of the rod's action, a conception that probably received an impetus from the greater knowledge of electrical currents that accrued from the labours of Volta and his successors.

At the present time the lightning rod is generally held by physicists to possess two functions; one, by means of the root and point, tending to prevent thunderbolts; and the other, through the medium of the stalk, tending to facilitate their passage after they had occurred, so that they should effect only a minimum of injury.[2]

The latter function is that which has been most readily adopted, especially in England, where probably a proportion of quite nineteen persons out of twenty are ignorant of any other.

The general supposition in this country appears to be that Franklin with his kite drew lightning from the skies,[3] and by thus producing it (as it were) in a tame and controllable form, was enabled to identify it with the electric spark; and that then, by analogy of reasoning, he conceived the idea of fixing metal rods to buildings in order that, if the lightning struck them, it might be drawn down the rods as it had been drawn down the wet cord of his kite.[4]

What really happened, however, in the case of Franklin's kite was this:—During the progress of a thunderstorm, the point that surmounted the kite elicited electricity from the ground in the visible form of an electric spark leaping across an air gap of slight length purposely left in the connection of the kite apparatus with the earth.[5]

This apparatus was simply a species of lightning rod, and the great elevation given to its point by the kite

---

[1] II. A 15; C 14, 42.
[2] II. C 42—45; G 31, 44, 48. VI. B a.
[3] II. A 20; C 42. VI. B a.
[4] II. A 14.   [5] II. A 14.

VIII. A b.

arrangement was quite unnecessary, as was proved by D'Alibard, who, at Franklin's instigation, produced the same results by means of a pointed iron rod of no great height.[1]

It has already been explained that an electric spark across a gap in a conductor is an explosion of a condenser temporarily formed by the gap and the ends of the conductor at the sides of the gap;[2] and, in accordance with this view, the reason for the sparks and flashes, drawn by Franklin, D'Alibard, De Romas, and others, is as follows:—

The charge on the surface of the ground was condensed by the thundercloud to a high state of potential; consequently, the lower part of the apparatus electrically connected to the ground shared this state.

Charge of an opposite nature was then induced, across the air gap, on the lower end of the disconnected upper part of the apparatus, the complementary electricity being driven upwards to the point, and then (assisted by induction from the cloud) ejected thereat.

Almost as fast as explosions in the shape of sparks occurred across the air gap, fresh charge accrued from the ground, and fresh electricity of the same nature was driven out at the point; and thus, through the leaking action of the point,[3] which allowed room for the induced charge above the gap to accumulate to explosive potential, a succession of sparks was produced during the period that the thundercloud remained within condensing range.[4]

Eripuit fulmen cœlo is therefore merely a figure of speech;[5] for what Franklin saw was not lightning; nor has any one ever intentionally drawn lightning from heaven, or created thunderbolts. Franklin's experiment only tended to confirm the preconceived idea that lightning was associated with electricity, and was probably a huge electric spark;[6] and it proved that pointed metal rods had

[1] II. A 13.
[2] VI. B b.
[3] II. C 53—55.
[4] II. A 13, 14, 22, 23.
[5] II. A 20.
[6] II. A 6, 7, 9, 12.

VIII. A c.

the power of ejecting electricity from the ground, at all events whilst thunderstorms were in progress, and the result was the conception of defending buildings by means of these rods.

There is hardly any doubt that the supposition of lightning passing down lightning rods into the ground arose from the prevalent view that its direction was invariably from the clouds to the earth.[1]

That Franklin himself gradually fell in with this theory of the function of lightning rods is almost certain; and the probability is that, after he had established the practice of his invention, he had neither inclination nor time to attempt to elaborate its theory.

It is not, then, to be wondered at that, in England, the term "lightning conductor" soon superseded that of lightning rod, and that the function of the rod's point was, for a long time, almost ignored;[2] although, inasmuch as the existence of lightning depends on the absence of conducting bodies, the expression "lightning conductor" clearly constitutes a contradiction in terms.

On the Continent, the question of lightning rods in general, and of the preventive power of their points in particular, appears always to have received more attention than in this country:[3] it was from France that the knowledge of Franklin's invention spread into England;[4] and it is worthy of notice that the term *paratonnerre*, used by the French to indicate a lightning rod, much more clearly denotes its preventive action than the corresponding English term.

(c) *The Opposition to the Use of Rods.*

It would appear that when lightning rods were first made known, and for some three-quarters of a century after-

---

[1] VI. B a.
[2] II. C 12, 35; D 33, 38.
[3] I. G 38; II. G 34.
[4] II. A 12.

VIII. A *c.*

wards, much opposition to their use was manifested, and even among men of known scientific attainments.[1]

Attempts have been made to show that this opposition was based on ignorance; but it is unquestionable that these opponents of rods had solid facts to go upon,[2] however inaccurate may have been their explanation of these facts.

Lightning rods, and the buildings they were intended to protect, have unquestionably often been the scenes of explosions; and in many cases it could not have been unreasonable to have treated the rods themselves as the causes of these disasters.

At all events, there was the fact of the explosion; and it is only in the nature of things that the sufferers from the injuries caused by it should, not infrequently, have considered that any explanation of it, however scientific, and however much the rods were exonerated, was but a poor consolation to them, should have declined to have anything more to do with such contrivances, and should have advised their acquaintances to act in the same way.

The knowledge of rods being frequently struck is probably the reason why, in this practical country, they are by no means universally adopted on buildings which present elevated features that might reasonably be considered to entitle them to the chance of obtaining whatever advantage might be desirable from the employment of such apparatus, *e.g.* country churches with towers, many of which are without rods.

The conflicts of opinion amongst the advocates of rods[3] may probably have also contributed to the development of the general indifference on the subject which has existed amongst Englishmen generally, but especially amongst English architects.

[1] I. E 1. II. A 17; G 3, 4, 7, 8, 46.
[2] II. A 16, 27, 33—35; G 6. VIII. C 1, 2.
[3] I. C [46, 48]; E [4, 9]; [E 1, II. G 15]; II. B [20, 27], [31, 53], [39, 60]; C [17, 35], [29, 46]; [C 34, E 12]; [D 22, G 49]; E [5, 40], [29, 38]; [E 16, G 45]; E 41; G [26, 39], [16, 55], [42, 54]; G 36; [G 32, B 46].

VIII. A d.

(d) *Diverse Systems of Application.*

At the present time there are two schools amongst lightning authorities.[1]

The chief school, comprising the great majority of physicists and engineers, adopts the principle, prominently expressed in Sir William Snow Harris's works, of embracing the building to be protected in a network of metal so as to bring it as nearly as possible into the condition of a metal building.[2]

In this system, the stalk of the rod is metallically connected to all the metal surfaces on the building, and is moreover supplied with branches in the form of metal bands which proceed along the roof ridges, gable ends, eaves, valleys, and salient outlines of the upper part of the building generally.[3]

The stalk is thus only the backbone of the protective system of the building; and the object of this system is, through the conductive power of the metal, to afford an easy passage to the root of the rod (*i.e.* to earth) for a lightning discharge that may strike any part of the building from above.[4]

During the last few years this system has received further elaboration from eminent electricians by an increase of metal both as regards the number of points, and also as regards the extent and ramification of the roots.[5]

The plan recommended and practised by the school of the minority is to place the rod well clear of any existing metallic surfaces, to insulate the stalk at the necessary points where it is fixed to the building, and to make its course to the earth as direct as possible.[6]

The metallic connections with the stalk, and the

---

[1] II. E 28, 29.   [2] II. E 42; G 19.
[3] II. B 5, 21, 40, 45, 56; E 13, 16—19, 25, 26, 28, 31, 36, 38; G 37, 45, 50. IV. 19.   [4] VI. B *a*.
[5] II. B 18, 29; C 19, 22, 36, 49; D 9, 12, 52, 54; G 19.
[6] IV. 1, 3, 28, 30, 31, 37, 39—41, 43, 45, 47, 48.

VIII. B.

branches therefrom, are held by these dissenters to be superfluous when the rod is in good order, and dangerous when it is not.[1]

(B.) COMMENTS ON THE THEORY AND PRACTICE OF LIGHTNING ROD DEFENCE.

The advantages afforded by a lightning rod appear to be summed up in the fact that, if it be in good condition, it tends by means of its point to furnish an outlet for any charge that may collect at the particular parts of the ground in contact with its root.

In the case of a building, if the surface of the ground around it have good collective qualities, and constitute one collector,[2] doubtless, if the root pierce it anywhere, it would tap the whole of it; and thus the rod would help greatly to defend the building from explosion.

Even then, however, the protective power of the rod is apt to be modified by reason of the fact that ordinary ground (moist earth) is not an accumulator of electricity like metal,[3] and that an appreciable amount of time must be required for the whole of a charged strip of ground around a building, or an elevated feature of a building, to empty itself through the rod; and unless the root completely permeates this ground and envelopes the building or its prominent feature, the potential of the charge of the ground at the parts of the base not thus embraced is liable to increase at a greater rate than that at which the charge is drawn away by the action of the point; and thus, in spite thereof, explosion may ensue.

The usual practice, however, is to lead the root of a rod directly *away* from the walls,[4] principally in order the better to obtain moisture;[5] hence it follows that the rod

---

[1] II. E 29.   [2] V. A 9.   [3] VI. D *a*.
[4] III. 56.   [5] II. D 1, 12, 13, 15, 20, 32, 35, 52, 54.

VIII. B a.

taps a considerable extent of ground which has nothing to do with the protection of the building.

In the case of a ship afloat, the iron or coppered bottom of the hull forms an excellent root, completely covering the ship's contact with the sea, and tapping the latter all round the ship's sides; and if the stalks running up her masts and rigging be efficiently connected to this root, as they usually are, the ship's system of defence, *quâ* her power of eliciting the earth's electricity that may collect around her, and thus of preventing explosion, is by no means a bad one, and probably greatly superior to the defence given by lightning rods to the great majority of buildings and constructions ashore.[1]

The principal disadvantages appertaining to lightning rods attached to buildings appear to be as follows, viz. :—

> (*a*) The exposure of elevated metallic surface furnished by them.
> (*b*) Their costliness.
> (*c*) The sources of failure to which, after erection, they are liable.
> (*d*) Their tendency to disfigure the appearance of buildings.

(*a*) *The Exposure of Elevated Metal.*

The elevation of the rod's point above the ground is probably due to some extent to the generally accepted maxim that a conical space, the extent of which depends on the height of the rod, is protected by it.[2]

When a rod is well connected to the ground it constitutes a local plate, and the exposure of elevated metal must always be a great disadvantage; for however well pointed this metal may be, and however thoroughly it may tap the ground, it must always be impossible to estimate whether

---

[1] II. A 30; B 6—15; G 23.   [2] II. G 1, 21, 35, 41, 47.

VIII. B *a*.

its tapping powers are on a par with its accumulative properties; and should the latter prevail, explosion will occur, and we can never foresee with certainty whether it will spring from the point (effecting, perhaps, no harm beyond blunting or fusing it), or from some lower portion of the stalk, or even from the whole of it at once; and it is clear that in the latter two cases the discharge may take a direction which may cause great injury to the building.[1]

Experience tells us that these "lateral discharges" or deviations of the explosion from the line of the rod are by no means rare occurrences.[2]

In cases where the piece of ground surrounding a building does not form one collective surface, but is broken up into several distinct collectors by means of intervening insulative or slowly collecting portions, it is evident that, under the present system of disposing a rod's root, there may be, close to the feature expressly intended to be defended, patches of surface which are not in any degree tapped by the rod.

Thus, in ground at no great distance from the rod a charge may accumulate and explosion may ensue; but now the rod's stalk, owing to its metallic substance, extent, and elevation, becomes very liable to be embraced in the explosion's line of least restraint, in which case the discharge probably leaps from the ground to the stalk over the surface of the building.

The stalk then, although itself a local plate, has accidentally acted as a local dielectric to the adjacent charged ground to which it is not connected, and, instead of having protected the building, has positively attracted danger to it; and this, although the apparatus may have been in a thoroughly efficient condition.[3]

When a rod does not "make good earth," *i.e.* when it is not electrically connected to the ground, it constitutes an

---

[1] VII. B *c, d;* G *e*. VIII. C 6.   [2] II. B 26, 46. VIII. C 11.
[3] VII. B *g*. VIII. C 7.

VIII. B *a*.

elevated local dielectric to all the ground below and around it; and experience fully shows that, in this condition, it is very liable to cause explosion and to be broken in pieces.[1]

Thus the stalk of a rod, whether acting as a local plate, or as a local dielectric, is a source of unmitigated danger; and this is in all cases considerably enhanced if it be formed of copper instead of iron; for the former accumulates electricity six times faster, and attracts explosion six times more powerfully than the latter.[2]

If the stalk of a rod be carried up inside a building, as it occasionally is to some extent,[3] a great part of the danger due to the exposure of metal would be obviated; but, on the other hand, there would be some risk, whilst the elevated pointed terminal remained outside, of the interior of the house being injured by explosions.

It would appear that Franklin's original intention actually was to carry up the stalks inside buildings, and that the practice was adopted on the Continent, but was soon abandoned.[4]

The fact of this abandonment, and of the almost universal system that has since remained in force of erecting rods outside buildings, seems to be a fair witness of an almost unanimous feeling that rods certainly attracted explosion however much they might afterwards annul its effects.[5]

The connections between the stalk and the metal surfaces on the building, as also the branches from the stalk along the ridges, eaves, gables, or other salient outlines of the building, of course add considerably to the amount of elevated metal exposed.

On board ship, the comparative danger from the stalk is probably less than on buildings, since the masts, yards, and rigging of ships always contain a considerable amount

VIII. B b.
of metal in an elevated position, and the extra metal due to the rod may therefore be relatively less felt.

(b) *The Costliness of Rods.*

The principal practical causes of expense in the first cost of a lightning rod appear to be four in number, viz. :—

(1.) The use of copper.
(2.) Expensive points.
(3.) The connections and branches of the stalk.
(4.) The extension of the root in search of moisture.

(1.) *The use of copper*[1] in lieu of iron, the material employed by Franklin,[2] appears to have been adopted mainly for the reason that, conductivity for conductivity, one is as cheap as the other, and that copper is the less corrodible of the two.[3]

The employment of copper rods has been mainly confined to these islands,[4] and seems to have been brought about chiefly through the influence of the late Sir William Snow Harris, to whom the metal recommended itself over iron, for employment on board H.M. ships.[5]

For the defence of ships the use of copper was gradually extended to that of powder magazines,[6] and other buildings.[7]

Copper may doubtless have been a convenient metal for adapting ships' masts to act as lightning rods, but there seems to have been no adequate scientific or practical reason for introducing the metal for the protection of constructions on land;[8] for iron rods, like any other forms of iron used on the exterior of buildings, could always have

---

[1] IV. 1—5, 8—11, 13, 18—26, 28, 29, 31, 33—36, 38—48.
[2] II. B 47.   [3] II. B 31, 35, 36.
[4] II. B 55, 57; C 50.   [5] II. B 5, 11—13. IV. 44.
[6] IV. 19.   [7] II. A 31; B 40.   [8] II. G 22.

VIII. B *b*.

been preserved from corrosion (before the galvanising process was known) by means of painting or tarring.[1]

The reception of the idea of using copper on buildings has led to unnecessary expense and danger; for the element of slenderness in rods of any material has generally been looked upon with disfavour;[2] and it is probable that in almost every instance where a copper rod has been employed, an iron one of the same dimensions would have answered the purpose equally well; whilst, *quâ* exposure of metal, it would unquestionably have been far less dangerous.[3]

Indeed, for all that is known to the contrary, an iron rod of the smallest practicable dimensions has a power for tapping any portion of earth it may be rooted in immensely in excess of what is needed to prevent that earth from accumulating potential.

Till good cause is shown why iron is inefficient for the purpose, the use of copper, independently of its intrinsic disadvantages,[4] would seem to be in the same category as that of silver and gold, and to be objectionable simply because it is expensive.[5]

(2.) *Expensive points.*—The prevalent employment on rods of pointed upper terminations made of copper, platina, or silver alloy, must of course constitute sources of expense.

The points of rods have been found to be liable to corrosion by the atmosphere and to fusion by lightning; and, as they are in such elevated positions that they cannot readily be visually examined or repaired, it has been thought that if they were formed of less corrodible or harder materials such as those mentioned, they would be less likely to need such examination and repair.[6]

The use, then, of these materials for the formation of

---

[1] II. B 55.
[2] II. B 5, 33, 40.
[3] II. B 27, 30; E 6.
[4] II. B 22, 35, 42, 53.
[5] II. A 31; B 50. VI. D *a*.
[6] II. C 10, 18, 20, 41, 47.

VIII. B b.

points is an expensive palliative rather than a sound remedy for the impossibility of constantly observing them and keeping them sharp.

Another costly arrangement is the brush or multiple point that is now frequently used.[1]

This brush is a cluster of three or more points, instead of a single one, at the end of the stalk; and the principal reasons for its use appear to be that there is a greater chance, among a number of points, of one remaining sharp, and that even if they all become blunted, still a number of blunt points is better than only one.[2]

Here the remedy appears to be not only imperfect and expensive, but also dangerous, for metal is thereby concentrated in the most elevated position.

(3.) *The metal bands forming connections* between the stalk and the metallic surfaces on the exterior of a building, and embracing its ridges, eaves, gables, and other prominent outlines, are clearly arrangements which, especially when formed of copper, increase considerably the cost of the rod.[3]

(4.) *The extension of the root* becomes a source of expense where, as frequently happens, there are no underground metal pipes near the building, and where there is no water or moisture in the vicinity.

It is customary to utilise as the roots or "earths" of lightning rods the gas or water mains of towns, on account of the large area of metal in contact with the ground afforded by these pipes, and of the facilities considered to be furnished thereby to the lightning discharge for dispersing itself harmlessly in the earth's reservoir.[4]

In the country the roots of rods are usually led into the the nearest well, sheet of water, or permanently moist ground.[5]

[1] IV. 3, 21, 29—32, 39, 41, 42, 47, 48.
[2] II. C 6—8, 19, 22, 32, 36, 49.   [3] II. E 36. IV. 19.
[4] II. D 1, 12, 13, 15, 32, 55.
[5] II. D 7, 11—13, 15, 21—23, 26, 27, 34, 36, 40, 43—45, 48, 49, 51—54.

VIII. B *b*.

If permanent moisture, however, should not exist near country buildings, the plan advocated is to enclose the roots within trenches filled with coke or charcoal embers, and to extend these trenches in every direction as far as practicable, so as to obtain as much temporary moisture as possible, and by means of a large area of root to compensate for the want of such a receptive state of the ground as alone is afforded by permanent moisture; moreover, rain water from the building is, where such a plan is feasible, to be led over the sites of these trenches.[1]

In the country, then, the roots of lightning rods, when arranged according to modern ideas, would frequently be costly; but it is obvious that, according to the theory on which they are disposed, they may also frequently be ineffective, however expensive and troublesome the laying of them proved to be; for buildings are constantly erected on dry and rocky sites so far removed from moisture that the excavations necessary to bring the roots thereto are out of the question.

In such cases, so far as can be seen, it is admitted by their advocates that lightning rods cannot be relied on.

Nothing can show more clearly how little the preventive power of the rod's root and point is thought of than the fact that such a practice as that of seeking moisture for the root is actually nearly everywhere in force.

If it were allowed that in protecting the building the function of the rod, or even one of its functions, was to tap and rid the ground of a certain element, it is inconceivable that the presence of ground full of the conditions for collecting this element should be considered a *sine quâ non* for the due protection of the building.[2]

If we wish to defend a place from an enemy we do not usually consider the presence of conditions favourable to an attack as essential to the due protection of the place.

On the other hand, if it be held that lightning descends

---

[1] II. D 2, 9, 10, 13, 15, 28, 31, 35, 46.   [2] VII. A *b, c*.

VIII. B *b*.

from heaven, strikes the rod, is conducted by it into the earth's reservoir, and needs a sympathetic or receptive state on the surface of that reservoir in order that it may disappear therein quietly and without obstruction,[1] then the practice we have mentioned would seem strictly rational and suitable.

Apparently a lightning discharge is usually treated as if it were a telegraph current, which, as we know, in order to complete its circuit without a return wire, is in need of "good earth."

The telegraph "earths" are accordingly either formed of gas or water mains, or of metal plates buried in earth which is either naturally, or artificially, moist.[2]

It seems to have been considered that an analogous course of treatment was necessary for a lightning discharge,[3] in order, apparently, that it should complete its circuit, *i.e.* vanish, as quickly as possible.

It is only within the last thirty years that a practice, previously in vogue, of constructing small masonry tanks, kept full of water, to receive the roots of rods, has been discontinued :[4] what the function of the rod's point was considered to be under these circumstances it is difficult to imagine.

We consider, then, that the attachment of rods to gas and water mains is simply giving these mains an additional, and generally unnecessary, outlet for the electricity tapped by them throughout the town ; and that the building against which the rod thus rooted is fixed is more exposed to injury from lightning than the other buildings of the town.

And, in the case of a rod whose root is laid in rivers, or in wells, or in water or moisture of any kind, not immediately contiguous to the construction, we maintain that the latter, so far from being protected, is merely being utilised to tap

---

[1] II. G 48.   [2] II. D 42.   [3] VI. B *a*.
[4] II. D 3, 33, 37, 38.   III. 43, 44.   IV. 19.

VIII. B c.

the electricity of this water or moisture, and that the rod is a distinct source of danger to it.

(c) *The Sources of Failure to which Rods are liable.*

The rod having been erected according to the best scientific advice, then comes the question of the disadvantage due to the various elements in its construction which experience has proved are apt to render it a failure *quâ* the purpose for which it was designed.

The sources of failure inherent in rods after their construction, as stated by their advocates, are as follows, viz. :—

    (1.) The stalk being of insufficient sectional area.[1]
    (2.) Bad joints along the length of the stalk, or between the stalk and the root; or deficient connection between the stalk and the metals of the building.[2]
    (3.) Alterations made in the arrangement of the metals of a building after the rod has been erected.[3]
    (4.) Blunt points.[4]
    (5.) The root, or earth connection, when tested with a galvanometer, displaying too great a resistance.[5]

(1.) *As regards sectional area* the opinions of authorities are conflicting : some say that a rod of small sectional area has advantages, whilst others doubt if it is better than none at all.[6]

It is clearly a considerable disadvantage to the present system of rods that, on such an important question, doctors should disagree.

[1] VIII. C 5 α.   [2] II. F 6—8. VIII. C 5 β.
[3] VIII. C 5 γ.   [4] II. F 6; G 18.
[5] II. F 1—4. VIII. C 5 δ.
[6] II. B 16, 20, 23, 27—30; G 16, 17, 55.

VIII. B c.

(2.) *Bad joints* along the stalk's length are hardly so likely to occur in these days as of yore; for the kinds of stalks mainly used for the protection of buildings in the present day are wire ropes,[1] needing no joints such as those required in the solid bars, bands, tubes, and chains, which were formerly employed.[2]

The action of the wind on the more or less unsupported pointed terminals projecting above the building is, however, with all kinds of rods, very likely to work loose the connection between these terminals and the supported portion of the stalk below.[3]

The possible disconnection underground of the stalk from the root is a source of failure that seems always liable to occur on account of the electrolytic action set up partly by the constant flow of a current from the earth through the rod, owing to its tapping powers, and partly by the moisture in the ground: by this action the metal appears to be gradually eaten away near the surface.[4]

Another cause of disconnection between the stalk and the root happens when the former is of copper, and the latter (as it usually is) of iron: in this case galvanic action is apt to be set up by the contact of the two metals in the presence of moisture, and the iron is decomposed.[5]

Want of connection between the stalk and the metals of the building is a reason that has been advanced for the failure of rods.[6]

The remedies that have recently been proposed for the existence of bad joints and disconnections are the constant inspection of the rod, and the periodical testing of its continuity by means of a galvanometer.[7]

(3.) *Alterations made in the positions of the metals* of a

---

[1] IV. 1—5, 7, 8, 21, 22, 24, 28, 29, 31, 33—36, 39—43, 45—48. II. B 1, 15, 17, 27, 30, 44, 59.

[2] II. B 6—9, 12, 25, 37, 41, 48; F 7. IV. 6, 9—20, 22, 23, 25—27, 30, 32, 37, 38, 44.

[3] II. G 51; F 6.

[4] II. D 8, 16, 18, 19; F 8. IV. 25.

[5] II. B 38; D 30, 39.

[6] III. 129.

[7] II. F 1—5, 9—14.

VIII. B *c*.

building, after a rod has been erected, have been found to vitiate its usefulness; and it is quite conceivable that such metals have in some instances been placed in positions where they have formed, in conjunction with part of the stalk, lines of least restraint from the ground, thus causing the stalk to form an accidental local dielectric, and effecting injury to the building.[1]

It would appear necessary then that, at the proposed inspections of the rod, the positions with regard to it of all metals, both inside and outside the building, should be examined.

(4.) *Blunt points.*—The only effective remedy for the points becoming blunted would appear to be visual examination of them; but this would be quite impracticable in the case of church spires and tall chimney-stalks without the aid of scaffolding or other expensive appliances.

(5.) *Resistance of the root's earth contact.*—The last and most important of the sources of failure mentioned is that alleged to be due to the rod "making bad earth."

It is held by lightning engineers that a rod is untrustworthy, *i.e.* that it cannot be depended on to lead away harmlessly the lightning discharge striking it,[2] if, on its "earth" being tested by a battery current of certain strength, a resistance above a certain standard is recorded on the galvanometer.[3]

It is not stated what this standard is, nor how it has been, or can be, determined with accuracy;[4] but, allowing that this point can be, and has been, settled, it is evident that great expense may have been incurred, when the rod was erected, in bringing its root to what might (taking one year with another) be fairly reckoned as permanent moisture; yet, should drainage operations have taken place in the adjacent ground, or should a period of drought supervene, and should this moisture be reduced to such a degree that the

---

[1] II. G 53. III. 158.
[2] VI. B *a*.
[3] II. F 1, 4, 9, 14.
[4] II. F 15.

VIII. B d.

galvanometer standard of resistance would be exceeded, then apparently the rod would be useless until remedial measures of some kind had been taken.[1]

Supposing, however, that at a testing the resistance of the " earth " should be found to exceed the standard, the course to be pursued seems by no means clear. How are we to remedy the evil? Are we to open the ground over the course of the root, and to examine the latter visually to see if anything is at fault with it? Or are we to presume that the fault is altogether with the state of the ground; and if so, by what practical means are we to render it, *quâ* the object we have in hand, more moist?

Altogether, the rule laid down for bringing the roots of rods into moist earth, and for testing them periodically as to their efficiency, is apparently surrounded with so much difficulty and uncertainty that, alone, it would seem to constitute a fatal drawback to the employment of lightning rod defence in accordance with modern views.

Even if this system should result in obtaining a protective apparatus always to be relied on, the disadvantage due to the trouble and expense of the periodical testing by electricians would still remain.[2]

(d) *The Tendency of Rods to Disfigure Buildings.*

The minor disadvantage due to the tendency of rods to disfigure the appearance of constructions is one that chiefly applies to churches, edifices containing some architectural adornment, and monuments or columns; and it cannot be said to apply at all to such constructions as powder magazines and chimney-stalks.

The features to which rods are affixed are always the most prominent ones of the building; hence it is difficult to conceal the rod, and its visibility is enhanced when brush points or insulators are employed.

Owing, therefore, to the conspicuous positions they

---

[1] II. D 47, 49, 54.   [2] II. F 14.

VIII. C 1 a—γ.

occupy, and to their incongruity with their architectural surroundings, rods are apt to mar the appearances of buildings.[1]

Of course this disfigurement is reduced considerably when the stalk is carried up inside the building; but, as we have stated, this course is an unusual one, and, even when it is adopted, the 'pointed terminal may be a very unsightly object, as, *e.g.* in the case of the Duke of York's Column in London.

(C.) ANALYSIS OF LIGHTNING INCIDENTS CONNECTED WITH RODS.

Attention will now be drawn to such of the incidents mentioned in Chapter III. as record that lightning rods, or the constructions they were designed to protect, were struck by lightning, or that rods were instrumental in preventing explosions. (N.B.—The nos. marked (*s*) are cases of ships, and those marked *c* are instances of copper rods.)

(1.) RODS STRUCK. (45 CASES.)

(a) *No damage done.* (13 cases.)

Nos. 8, | 66, | 78, | 100 *c*, (*s*), | 120.
22, | 69 *c*, (*s*), | 79, | 101 *c*, (*s*),
56, | 70 *c*, (*s*), | 99 *c*, (*s*), | 102 *c*, (*s*),

(β) *Only the Rods damaged.* (11 cases.)

Nos. 5 (*s*), | 46, | 64 (*s*), | 92,
6 *c*, | 49, | 65 *c*, | 156 *c*.
44, | 59 *c*, | 89 *c*,

(γ) *Both Rods and Constructions damaged.* (15 cases.)

Nos. 19, | 47, | 86 *c*, | 108, | 150,
21, | 55 *c*, | 91 *c*, | 128 *c*, | 153,
43, | 58, | 103, | 131 *c*, | 179.

[1] IV. 1, 3, 6, 14, 15, 18, 24, 25, 27—29, 34, 41, 47.

VIII. C 1, 2—5 a—γ.

(δ) *Constructions damaged, but not the Rods.* (6 *cases.*)
Nos. 72 c, (s), 73, 104 c, (s), 114, 118 c, 157 c.

(2.) CONSTRUCTIONS STRUCK, BUT NOT THE RODS. (16 CASES.)

(a) *Constructions damaged.* (11 *cases.*)

Nos. 71 c, (s),   105,   130 c,   158,
74,   110,   147,   160.
75 (s),   127,   155,

(β) *Constructions damaged and persons injured.* (2 *cases.*)
Nos. 129 c, 134 c.

(γ) *Persons at or near the Constructions killed or injured.*
(3 *cases.*)
Nos. 45, 82, 115.

[NOTE.—In No. 82 the rod was probably struck, though the record specifies the magazine.]

(3.) CONSTRUCTIONS WITHOUT RODS, DAMAGED, CLOSE TO OTHER CONSTRUCTIONS WHICH HAD RODS, BUT RECEIVED NO INJURIES. (2 CASES.)
Nos. 48, 181.

(4.) BUILDINGS STRUCK BEFORE BEING SUPPLIED WITH RODS, BUT NOT SUBSEQUENTLY. (4 CASES.)
Nos. 57, 77, 80, 81.

(5.) CAUSES TO WHICH THE FAILURES OF RODS HAVE BEEN ATTRIBUTED BY AUTHORITIES.

(a) *To insufficient sectional area of Rod.* (10 *cases.*)

Nos. 5,   56,   59,   65,   131,
19,   58,   64,   118,   134.

(β) *To bad joints or deficient connections.* (4 *cases.*)
Nos. 5, 114, 129, 131.

(γ) *To the metals of the building.* (2 *cases.*)
Nos. 45, 158.

VIII. C 5 δ, ε; 6—10.

(δ) *To bad earth.* (17 cases.)

Nos. 43,	86,	118,	131,	156,
44,	89,	127,	134,	157,
55,	91,	128,	155,	179.
64,	115,			

(ε) *To the Rods being "not so perfectly applied as they ought to have been."* (1 case.)

No. 73.

(6.) RODS WHICH PROBABLY ACTED AS LOCAL PLATES. (22 CASES.)

Nos. 6,	66,	72,	79,	99,	102,	120,
8,	69,	73,	82,	100,	104,	150,
21,	70,	78,	92,	101,	108,	153.
49,						

(7.) RODS WHICH PROBABLY ACTED AS ACCIDENTAL DIELECTRICS. (9 CASES.)

| Nos. 5, | 22, | 56, | 59, | 114. |
| 19, | 47, | 58, | 65, | |

(8.) RODS WHICH PROBABLY ACTED AS LOCAL DIELECTRICS. (15 CASES.)

Nos. 43,	55,	89,	118,	156,
44,	64,	91,	128,	157,
46,	86,	103,	131,	179.

(9.) ENDS OF RODS FUSED.

Nos. 5,	55,	128,
6,	64,	150,
21,	92,	153,
46,	103,	156.
49,		

(10.) MECHANICAL INJURIES TO RODS.

Nos. 44,	58,	92,
46,	59,	108,
47,	65,	131,
49,	86,	179.
55,	89,	

VIII. C 11, 12.

(11.) DEVIATIONS OF THE EXPLOSIONS FROM THE RODS.

Nos. 43,	114,
55,	128,
58,	131,
86,	150,
91,	153,
103,	179.

(12.) NOTES.

In regard to the 13 cases of Group (1), in which the rods were struck, but it so happened that no damage was occasioned to the constructions, if the theory be sound that a rod is placed on a building in order to waylay the lightning in its supposed descent,[1] and to afford it an easy channel to the earth,[2] these incidents can only be looked upon as illustrations of the success of the rods.

But if the theory which we have striven to prove be correct, and if the true function of lightning rods be to prevent explosions occurring at all,[3] then unquestionably these 13 cases, equally with the 32 others of Group (1), in which the rods were struck, and damage also was done, are clear instances of failure; for every one of these 45 instances is a proof that, instead of preventing explosion, the rod caused it.

Let us put the matter in another form. If the rods had no share in causing these explosions, why were they struck?[4] Is it not manifestly in accordance with electrical law that no single portion of an exploding condenser can exist that has not had its share in determining explosion?[5]

If the views we have upheld be sound, it is clear that nothing in nature can be struck by lightning without having first caused lightning to strike it.

---

[1] VI. B a.
[2] II. G 5, 6, 31, 39, 48.
[3] II. C 30; G 40, 56; E 27.
[4] II. G 6.
[5] V. B 11.

VIII. C.

In the 16 incidents of Group (2), where the rods were not struck, but the buildings or persons in them were injured (the special cases of Nos. 74 and 115 excepted), the rods, according to the generally received views, must be considered as having failed to exercise their supposed protective influence; but if they be viewed in the light in which we have been treating them, they cannot be altogether treated as failures; for they evidently protected the building to the best of their ability by preventing explosions occurring at the places tapped by their roots; whence it followed that they themselves were not struck, and that the scenes of the explosions were distant from them and probably outside the influence of their roots.

Here, undoubtedly, the beneficial power of the roots and points overbore the baneful accumulative action of the stalks; and the rods were successful so far as the disposition of their roots permitted.

Out of the 61 cases comprised in Groups (1) and (2), the following, by reason either of their instructive character or of the fulness of the information that has been procurable concerning them, appear to be well worthy of attention, viz. :—

Buildings.

No. 43. Milan Cathedral.
„ 44. Genoa Lighthouse.
„ 47. Bayonne Powder Magazine.
„ 49. Signor Melloni's house.
„ 55. Normanhurst Court.
„ 56. The Sorbonne.
„ 82. Glogan Powder Magazine.
„ 86. Warehouse near Victoria Station.
„ 114. Barrackpore Barracks.
„ 115. Buys' house at Natal.
„ 118. All Saints' Church, Nottingham.
„ 128. Rostall Church.
„ 129. Bruntcliffe Colliery Powder Store.
„ 130. Cromer Church.
„ 131. Laughton Church.
„ 134. Wrexham Church.

No. 155. St. Mary's Church, Genoa.
„ 156. Alatri Cathedral.
„ 157. Clevedon Church.
„ 160. Compton Lodge, Jamaica.
„ 179. East London Powder Magazine.

Ships.

No. 5. Packet Ship *New York*.
„ 65. H.M.S. *Dublin*.
„ 69. H.M.S. *Beagle*.
„ 71. H.M.S. *Endymion*.
„ 72. H.M.S. *Etna*.
„ 104. H.M.S. *Racer*.

Of the above cases, Nos. 49 and 56 are very remarkable, as clearly showing the power of rods to attract lightning; and especially No. 49, which Arago has recorded with some circumstantiality.

Another noteworthy case is No. 156; firstly, because the incident narrated is shown to be the fifth occasion since the rods were erected, *i.e.* in eight years, that one or other of them had been struck; secondly, because, in the case in question, the attractive power of the rod is shown by the fact that the explosion actually cut its way horizontally through 11 yards of hard ground in order to get at it; thirdly, because the root appears to have shared with the stalk its function of local dielectric. The damage at the Alatri Fountain was probably caused by a simultaneous explosion thereat.

The cases in Group (3) are to some extent evidences of the usefulness of rods; for in these cases explosions happened at constructions not defended by rods, close to a number of others which were so protected; but none of these protected buildings or their rods were struck.

The facts related in the 4 cases that constitute Group (4) are important, as showing that the buildings in question had been repeatedly struck before rods were applied, but never afterwards; and this would seem to testify to the benefit that arose from their presence.

VIII. D.

D. Summary of Remarks on Lightning Rods.

The advocates of rods in their present form appear to maintain that any danger due to the attractive influence of the stalk is practically neutralised by its conducting power, and that these characteristics do not hinder the rod from faithfully performing its preventive duty: in other words, they hold that the rod, through its root and point, unceasingly tries to prevent lightning from coming to the building, and that if it should herein be unsuccessful, its stalk becomes useful, [inasmuch as it furnishes, firstly, a prominently attractive object by which the descending lightning[1] is intercepted from striking the building, and, secondly, a good conducting channel by which the discharge is led harmlessly into the ground.[2]

But if we once entertain the notion that lightning springs from the ground[3] and cannot be intercepted or conducted, the *raison d'être* of the stalk ceases, and our course becomes limited to preventing the development of those conditions which go to form lightning; and, according to all experience, and to what is now admitted by lightning engineers, stalks and elevated metals of any shape on the exterior surfaces of buildings are certainly among these conditions.[4]

According to our views, then, there is no sound foundation for the existence of the stalks of lightning rods, and their presence is an unmixed source of danger and expense.

It may be urged by the advocates of lightning rods that, during the one hundred and thirty years that have elapsed since they were invented, the vast majority of buildings to which they have been affixed have not been harmed by lightning, and that they can hardly be expected to abandon this time-honoured system unless some practical proof of a better one be given.

---

[1] VI. B *a*.   [2] II. C 43—45; G 44.   [3] VI. B *b*.
[4] II. G 14, 15, 25—27. I. E 9.

VIII. D.

In reply, however, it may be asked, would these buildings ever have been injured if the rods had not been present?

It may even be asked, how can we know that these buildings did not receive their immunity in spite of their rods?

As regards practical proof, it is obvious that, until we are able to forge thunderbolts, the efficacy of any defensive measures devised to prevent them cannot be tested.

Our guide, then, in the matter of lightning engineering must still be, as it always has been, theory. Now we are told on good authority that theory is "a sort of intellectual contrivance for representing to the mind the order and connection subsisting between observed phenomena."[1]

The marshalling of observed actions and incidents in connection with lightning is, therefore, what we must rely on as the basis of our system of protective measures.

In conclusion, we may sum up our opinion on the present system of lightning rods as follows:—

> (1.) Owing to their roots and points they have a tendency to protect the constructions (and especially the ships) to which they are affixed; and they frequently have done so.[2]
>
> (2.) Owing to their stalks they have a tendency to injure the constructions to which they are attached; and they constantly have done so.[3]
>
> (3.) When applied to the defence of buildings their roots are unscientifically arranged.[4]
>
> (4.) Their costliness* and uncertainty act as deterrents to the adoption of measures for obtaining defence from lightning at all,[5] and are especially injurious to the interests of the poorer classes.

[1] I. A 3. [2] II. B 13; G 2, 10, 20, 23, 34.
[3] VIII. A c. [4] VIII. B. [5] VIII. B, b, c.

* "Sir William Thomson, at the meeting of the British Association at Aberdeen, said, 'If I urge on Glasgow manufacturers to put up lightning rods, they say it is cheaper to insure than to do so.'"—*Report of Lightning Rod Conference.*

Part III.

PRACTICAL MEASURES ADVOCATED FOR THE DEFENCE OF LIFE AND PROPERTY FROM THE EFFECTS OF LIGHTNING.

CHAPTER IX.—PRACTICAL MEASURES ADVOCATED.

We have now arrived at that part of our treatise where it becomes necessary to show how the theories we have advanced can be put into practical shape.

The subject of the best means of defending life and property from the action of lightning appears to divide itself into three distinct phases, viz. :—

(A.) The defence of large areas, such as countries, districts, and towns.

(B.) The defence of constructions, *e.g.* buildings, mines, and ships.

(C.) The defence of individuals.

The great end of any defensive measures should undoubtedly be to save life; and the defence of buildings and ships is of course a valuable means to that end; but the protection of property, *quâ* property, is also necessarily of much importance.

It will be evident from the views upheld in Part II. that the great principle which we advocate to be carried out to the utmost extent in the defensive measures to be adopted in dealing with thunderbolts, is that of prevention.

IX. A a.

The three phases which we have enumerated will now be dealt with separately.

(A.)—THE DEFENCE OF LARGE AREAS.

(a) *The Defence of Countries and Districts.*

It is evident that any action taken with a view of preventing or diminishing the occurrence of thunderbolts in whole regions cannot well be undertaken by any other agency than the Government of the country.

The first step in such a course of action would appear to be the scientific investigation of each separate instance of death or injury occasioned by lightning, and the accurate record of the results of such investigations.

Thus each incident injurious to the community inflicted by thunderbolts would form the subject of Government inspection and report, on a system somewhat analogous to that now in force in England for inquiring into the causes of railway accidents, or of those due to the manufacture, storage, and transport of combustible materials.

So soon as facts had accumulated to a degree to warrant the assumption of some laws as to the manner in which, in the country in question, thunderbolts appeared to guide their action, then would probably be the time for the Government to consider the advisability of some definite course of operation as regards either the whole country, or any particular district or province whose circumstances would appear most to require such treatment.

It is impossible to forecast what direction this more detailed action of Government might take; but it is quite conceivable that great public benefit would ensue to certain localities from the trial of some inexpensive general measures which the experience resulting from the records of lightning incidents might show to be worthy of experiment.

No kind of lightning apparatus (however simple) for the defence of individual buildings could possibly compare in advantages with the adoption of a policy on the part of

IX. A *a.*

the authorities calculated to prevent, so far as possible, the occurrence of thunderbolts within the whole region; and by no surer method could life and property be more effectually preserved.

We believe that the amount of light that would be thrown on the action of lightning from the mere scientific inspection and record by authority, for a few years, of the thunderbolt incidents that should take place during that time would be something incredible; and if these records were published annually, the public interest would be kept alive, and the endeavours of the Government would be assisted.

If, in connection with this investigation, steps were taken to mark in some permanent way the exact spots, whether at houses or on the open ground, where thunderbolt explosions had happened, probably much advantage would ensue; and especially would this course seem appropriate in cases where life had been lost.[1]

We know from experience, and from the manner in which explosions must necessarily be influenced by terraneous circumstances, that there is good reason for believing that where a thunderbolt has once occurred it is likely to occur again.[2]

Hence a mark of some kind would be not only valuable as a memorial of the past, but also instructive as a guide for the future; and if the researches of the Government should lead, as perhaps they might, to some kind of electro-geographical or electro-geological survey[3] of a district being commenced, these marks would probably be of much assistance.

Again, a great deal of scientific information might probably be amassed through the agency of the Government, in connection with the proposed investigation, not only on the subject of thunderstorms, but also on that of all terrestrial phenomena more or less associated with electrical

IX. A b.

action, viz. earth currents, auroræ, magnetic storms, earthquakes, volcanic eruptions, and waterspouts.[1]

At the present time there appears to be no organization in the United Kingdom for obtaining and recording accurate information regarding accidents by lightning; and for whatever knowledge we obtain of them we appear to be mainly indebted to the pens of the gentlemen whose business it is to supply the local press with news of abnormal events,[2] for the metropolitan press does not seem to be able to devote much space to the publication of intelligence regarding the ravages of lightning.[3]

Under any circumstances the knowledge afforded by the press is but slight, and the death of an agricultural labourer by lightning elicits less publicity than would be the case if he had died from hydrophobia, or had been killed in a railway accident, or had been driven over in the streets; yet it is clear that in each instance of death by lightning immense benefit would be derived from a publication of the details of the scene of the occurrence, of the clothes and metals on the person of the deceased, of the traces left by the thunderbolt, and of the proceedings at the coroner's inquest.[4]

Of course the reason for the scanty information obtainable on this subject through the usual channels can be none other than the want of public interest in the matter; and if any measures on a comprehensive scale are to be adopted in this country with a view to the prevention of loss of life by lightning, it seems essential that the initiative should be taken by Government.

(b) *The Defence of Towns.*

The defence of towns from thunderbolts might doubtless be dealt with by the municipal authorities.

We have already suggested that underground metal pipes (*e.g.* gas and water mains) and pavements are

[1] I. C 73. [2] III. 137, 141, 146, 148, 149, 152, 173, 183—190.
[3] III. 151. [4] I. D 65.

IX. B.

probably elements of efficacy in preventing or diminishing thunderbolts.[1]

If these views, then, be sound, it would be a matter of consideration for the smaller towns, as yet unprovided with such advantages, to hasten the time for supplying themselves therewith.

Towns appear to have as a rule so much immunity from injury by lightning as compared with villages and the open country,[2] that probably no other protective measures than those above mentioned would be needed by them apart from the special defence of particular buildings containing prominent elevations.

(B.)—THE DEFENCE OF CONSTRUCTIONS.

Pending the gathering of more detailed and more scientific information regarding the action of lightning than we at present possess, the following suggestions are offered regarding the defence of constructions.

There appear to be two broad principles which may be taken as the basis of means intended to secure buildings from thunderbolt explosions, viz. :—

(1.) Insulation.
(2.) Leakage.

The insulating principle can be carried out by two distinct measures, viz. :—

- (*a*) The removal of metals and explosive conditions from the outer walls.
- (*b*) The reduction of the explosive conditions of the surface of the ground immediately around the building.

The leakage principle can also be carried out by two separate measures, viz. :—

- (*c*) The conversion of chimney grates into electric taps.

---

[1] VII. G *d*. [2] I. G 35, 36.

IX. B a.

(d) The application of electric taps to the ground surrounding the building.

We have thus four protective arrangements, any of which, according to circumstances, can be used for the defence of buildings.

(a) *The Removal of Metals and Explosive Conditions from the Building.*

If the views that have been laid before the reader on the subject of the danger arising from the presence of metals on the roofs and outer walls of buildings, and especially on their exterior surfaces, be correct,[1] it is manifest that in designing a building, the defence of which from thunderbolts happens to be an important consideration, care should be taken to avoid altogether the use of this material externally in positions elevated above the ground, and, as far as possible, internally, in contact with the roof and outer walls.

We consider that this absence of elevated metal from the outer walls is, with all buildings in special need of protection, a vital necessity; for (independently of theory) all experience points with unswerving finger to the law that exposed elevated metal is, as regards the action of lightning, a source of danger.[2]

We will now consider briefly some of the practical architectural arrangements which might be conveniently entertained at various kinds of buildings with a view to the avoidance of metal.

In the first place there are the factories, laboratories, and magazines of gunpowder and other combustibles;[3] these may be considered the forms of building which above all others require defence from thunderbolts.

In the design and construction of these buildings, there seems no reason why their exterior surfaces should afford

---

[1] VII. B *d, f, g, i*.   [2] C 3.   [3] VII. C 1, θ.

IX. B a.

the slightest appearance of metal; and, internally, the locks, bolts, and hinges of the doors and shutters would appear to be all that is necessary; for although the heavy vaulted roofs appropriate to powder magazines[1] are undesirable in factories and laboratories, yet very light Portland cement concrete flat arches (which are practically beams) might easily be adopted for these latter buildings.

As regards other kinds of buildings needing special protection from lightning, the following arrangements are suggested, viz. :—

>(1.) The material of the walls would be brick, stone, concrete, or other slowly collecting material, but not metal or wood.
>
>(2.) All metal work on the outside of the building, except close to the ground, would be rigidly omitted, *e.g.* weathercocks, spindles, finials, vanes, balls, crosses, chimney-pots, roof coverings, ridge crestings, eaves gutters, rain-water pipes and balconies.
>
>(3.) Wooden roof coverings would be avoided, and the use of woodwork on the exterior of the walls would be as sparingly adopted as practicable.[2]
>
>(4.) If ordinary pitched roofs be used, the covering would be of tiles in lieu of slates,[3] whereby the copper nails necessary for fixing the latter, and also the leaden flashings, ridge, hip, and valley pieces, frequently accompanying their use, would not be required; and the roof framing would be of wood, not of iron.
>
>(5.) Flat asphalted concrete fireproof roofs would (notwithstanding the iron girders buried in and supporting them) probably be fairly secure.
>
>(6.) Chimney-pots would be of earthenware or terra cotta.
>
>(7.) The terminations of church spires, pinnacles,

---

[1] IV. 19.   [2] VII. C 6.   [3] I. D 49.

IX. B a.

gables, &c., would consist of crosses, bosses, or other ornaments, made of stone or terra cotta.

(8.) Eaves gutters and rain-water pipes would be formed of stone, concrete, earthenware, terra cotta, or kindred materials, but never of metal or wood.

(9.) In the construction of all outer walls, the use of tie rods, hoop-iron bond, lead joints, gratings, and metal work generally, would be avoided as far as practicable; in block stonework in spires, towers, and elevated features, slate dowels would supplant iron ones, and the use of metal cramps would be avoided.

(10.) Clocks would not be set high up in church towers, turrets, campaniles, or other elevated features of buildings.

(11.) Bells would also not be set high up in church towers or campaniles.

(12.) Iron floors, columns, and staircases in the interior of the building would be avoided.

(13.) No masses of metal, *e.g.* safes, organs, and large mirrors, would be placed inside the building in contact with the outer walls; and smaller metal surfaces, *e.g.* gas and water pipes, would also be kept, so far as practicable, away from these walls.

(14.) Stained windows of churches[1] would not be covered with wire guards; probably stout glass shields could be employed as substitutes.

Probably other means would also occur to architects whereby the use of metal in and about the building might be diminished.

The foregoing proposals are not made with reference to all buildings, but only as regards those which in the opinion of the architect, owner, or occupier, demand special

---

[1] III. 111, 172.

IX. B *a*.

protection from lightning, whatever may be the cause assigned.

It is submitted that this category should generally include the following, viz.—all buildings connected with the manufacture, manipulation, or storage of powder or other combustibles; tall chimney-shafts; monuments and works of art; churches, and other buildings, whether in town or country, having elevated features, *e.g.* spires, pinnacled towers, domes, cupolas, turrets, and belfries; large stores, warehouses, and factories, in exposed positions; and country labourers' cottages.

These buildings would all be dealt with according to their respective circumstances so as to reduce the amount of metal about them to a practicable minimum.

The chimney-stalks of furnaces are, owing to their great height, and to the facilitative lining of their interiors, peculiarly liable to be struck by explosions; hence every precaution should be taken, in their construction, as regards metals.[1]

The custom of erecting stone columns of considerable height to support statues and other metal works of art[2] appears to have died out in this country; but, independently of the danger from being struck by lightning, it seems to be somewhat inappropriate to place statues of eminent men on pedestals so high that their features cannot be distinguished, besides being a very expensive course to adopt.

The vanes and weathercocks of gilt copper or other metal, which almost invariably surmount the spires and pinnacles of churches, seem to be specially useless and dangerous ornaments.[3]

When untarnished, the gilding has not an unpleasing effect to the eye, but it is hardly doubtful that delicately-carved Latin or Greek crosses or other stone ornaments would have an equally good appearance, and would be more in harmony with good architecture.

[1] III. 96. VII. B *k*.   [2] III. 74.
[3] III. 7, 51, 52, 62, 68, 131—133, 188.

IX. B a.

It is probable that metal ornaments on the summits of church spires have many a disastrous explosion to account for; take for instance the case of St. Martin's Church on the 28th July, 1842.[1] Here the vane was 8 feet long by 6 feet wide, and the vane spindle was 4½ inches square in section, and 27 feet long.

As regards church clocks and bells,[2] it is suggested that both might conveniently be placed much nearer the ground than is usually the case; that in many churches they might without inconvenience be placed in some part of the building other than the tower or most elevated feature;[3] and that, in some instances, they might advantageously be dispensed with altogether, as, in fact, in dissenting churches in this country is always the case as regards bells, and usually as regards clocks.

Large stores, warehouses, and factories, in exposed positions would probably receive advantage from such of the proposed measures regarding the removal of metal as could be conveniently applied to them.[4]

Labourers' cottages in the country, whether in villages, or isolated, are specially in need of precautions regarding metals;[5] for they appear to be visited by thunderbolts with a proportionally greater frequency than well-built houses.

Whether castles, halls, and gentlemen's houses in the country,[6] substantial farm-houses,[7] and farm buildings[8] should be treated in the manner proposed, or should be specially dealt with at all *quâ* defence from lightning, would depend a good deal on the circumstances of the locality, of the building, and of the occupier.

Some districts are seldom visited by thunderbolts;[9] and

---

[1] III. 51.
[2] III. 16, 36, 51, 52, 78, 79, 111, 128.
[3] I. F 21; G 13.
[4] III. 39, 86, 141, 189, 197.
[5] VII. C 1η.
[6] III. 49, 55, 67, 79, 117, 121, 160, 177, 183, 186.
[7] III. 184, 200, 201.   [8] I. G 31.   III. 113, 163, 202.
[9] I. G 11, 26, 27, 30, 32.

IX. B b.

in these regions there would be less necessity for taking special measures for the defence of private buildings.

The surroundings of the house, *e.g.* water, rock, trees, adjacent buildings, the features of the country, the elevation, the degree of exposure, are, of course, all matters to be taken into account.[1]

Lastly, the occupier's pocket, and his or her views generally concerning the importance of the dangers due to thunderbolts, are important factors in the question.[2]

It is considered that ordinary town houses, without prominent features, in towns supplied with gas or water, would not as a rule need special treatment for defence from lightning, either as regards metals, or in any other way.[3]

(b) *The Reduction of the Explosiveness of the Ground.*

It would be advisable to build all constructions on thoroughly dry sites, and, wherever practicable, on rocky ones.[4]

Powder magazines should be constructed underground whenever such a course is feasible.

All buildings, as to which it is an object that they should be specially protected from lightning, should be kept well away from the banks of small rivers and small sheets or surfaces of water;[5] and probably the sites of all new constructions of the kind we have proposed to take precautions with as regards metal, would be worthy of special attention in regard to the nature of the soil and of the surroundings.

These measures would, however, be hardly needed in the cases of such buildings forming parts of the streets of towns supplied with gas or water.

Having done, then, all we can, by natural means, to diminish in the immediate vicinity of such buildings as we

---

[1] VII. A. [2] III. 121. [3] VII. G d.
[4] VII. A d. [5] VII. A b.

IX. B *b.*

specially desire to protect from thunderbolts the conditions tending to cause them, the question arises whether the collectivity of the surface of the ground lying close to the walls may not be advantageously reduced by artificial means, viz. by paving.[1]

The constructions to which, provided they contained no metal on the exterior surfaces of their outer walls, the measure of paving the ground immediately around those walls might advantageously be applied, would probably be all those proposed to be treated in regard to metal, except labourers' cottages, where, in most cases, the expense would probably be too great.

The pavement proposed is one of the following kinds, or of a nature kindred thereto, viz. :—

 (1.) Squared stone flagging not less than $2\frac{1}{2}$ inches thick, set in cement, and bedded on concrete.
 (2.) Bricks, flat or on edge, set in cement, and bedded on concrete.
 (3.) Cement concrete.
 (4.) Concrete coated with asphalte.
 (5.) Tar pavement.

The total depth of masonry should in no case be less than 6 inches; the kerb should be at least 6 inches high above the ground adjacent; the width of the paving should be from 3 to 6 feet, and there should be thoroughly good contact between it and the wall of the building.

From a sanitary and a convenient point of view, a pavement around any building is always an advantage; and, *quâ* lightning, the walls would certainly when thus girt appear to run less chance of being utilised as a pathway by an explosion from the adjacent ground.

The system would be especially applicable to buildings of no great area, such as usually are magazines and factories for explosives, and monuments or works of art;

---

[1] VII. A *c.*

IX. B c.

for the expense of completely surrounding such constructions with pavement would not be great.

At buildings containing elevated features, as, *e.g.* furnaces with tall chimney-stalks, and churches with towers or spires, the pavement would, where desirable, merely surround the bases of such features, or would cover the nearest ground thereto close to the walls, if such bases did not reach the external ground. In the latter case, the following would be a convenient rule for obtaining approximately the length of pavement required, viz. find from the plan the girth which the base of the elevated feature would approximately have, supposing its walls or sides had everywhere reached the ground, perpendicularly from where they are stopped; and take this girth as the length of the inside edge of the necessary paving.

We have already adverted to the fact of most buildings in towns being enclosed by pavements.[1]

(c) The Conversion of Chimney-grates into Electric Taps.

The best forms of lightning protective apparatus for buildings now demand our attention.

In the country, the portions of houses that most cause explosions and most suffer from them are generally the chimneys.[2]

We propose, therefore, to apply a special arrangement to the fireplaces on the lowest floors of country houses and cottages, whereby, if possible, these masses of metal, and the sooty chimney flues leading from them, may have their usual function of local dielectric converted into the much less dangerous one of local plate,[3] and the grates or ranges themselves may form portions of taps for the purpose of ridding the ground adjacent to them of charge.

The proposed plan is merely to connect the grate by means of one or more iron bars to the ground below, and to fix on the grate a few short sharp iron spikes.

---

[1] VII. A *e*.   [2] VII. B *k*.   [3] II. E 7.

IX. B c.

The connection might be in the form of a single piece of iron of any convenient shape, riveted or otherwise fastened to the bottom of the ironwork of the grate, or range, and extending about 12 inches vertically into the natural soil below the hearth; but two or three of such connections would be better than one.

The spikes might be quite small (say 2 inches long), and might be fixed in any convenient parts of the grate, and so as to point either upwards, downwards, or in any other direction; for their action would be the same in all cases, and however placed they would certainly tend to eject at all times any electricity which might collect in the ground adjacent to the chimney.

The fact of these points being practically inside the house would not affect their power.[1]

The application of the above measure in its simplest form to the grate of a labourer's cottage in the country would be very inexpensive, and could easily be effected by a village blacksmith.

For the defence of the chimneys of sitting-rooms on the lowest floors of country houses it might perhaps be worth while expressly to manufacture grates with wide vertical horns projecting from their bottoms (to sink into the soil), and numerous ornamental pointed projections on any portion of them above ground.

The points of grates arranged as proposed could always be kept sharp without difficulty; and the hearth-stones could be arranged so that (if required) parts of them could be periodically lifted in order to examine the earth contact of the connections.

It is submitted that, with the basement fireplaces treated as proposed, labourers' cottages, and all ordinary houses in the country, such as, *e.g.* country gentlemen's seats, parsonages, and farm-houses, without prominently elevated features or any great amount of elevated metal on their

[1] V. B 20.

IX. B *d*.

exterior surfaces, and not otherwise in need of special treatment, would be fairly well defended from thunderbolts.

The chimney-stalks of furnaces [1] would also doubtless be rendered more secure if similar measures were adopted with regard to any masses of metal connected with the boilers or fireplaces lying near the bottom of the shaft.

(*d*) *The Application of Electric Taps to the Ground surrounding the Building.*

The form of lightning protector which we advocate as a substitute for the lightning rod in present use will now be described.

We have submitted that the advantages of the present form of lightning rod are measured by the extent to which, by means of its root and point, it taps the ground around the building; that as a rule the root is so arranged as to tap a minimum portion of this important ground; that the stalk is an unmitigated disadvantage; and that the system when carried out in the complete manner advocated by modern authorities is costly to erect, and liable to several sources of failure after erection.[2]

The principal conditions, then, that we propose to adopt in the new form of apparatus are as follows, viz. :—

> (1.) The protector must be arranged so as to tap the ground lying close around the building or feature to be defended to the maximum extent.
> (2.) It must have no stalk or exposed elevated surface of metal.
> (3.) It must be cheap to erect.
> (4.) It must not be liable after erection to any material source of failure.

Acting on these views, we suggest a form of lightning

---

[1] II. G 17.   [2] VIII. B D.

IX. B *d*.

protector for buildings which may not inaccurately be termed an electric tap.[1]

It consists, in its complete form, of a plate of iron laid in the ground a few inches below the surface surrounding the building, close either to the walls or to the kerb of any pavement, of the nature and dimensions proposed, adjoining them, and presenting slightly above the surface of the ground numerous sharp iron points.

Old metal of any kind or shape would do for the plate, and any sort of metal spikes attached thereto, and appearing above the ground just so high that the points could be conveniently kept sharp and free from injury, could act as points; but, in order to assist the ideas, it may be useful to suggest a definite form and dimensions, which, under ordinary circumstances, would be efficient, convenient, and inexpensive.

The following specification is therefore submitted for the construction and fixing of these taps, viz.—Wrought-iron plates, 4 feet long, 6 inches wide, and $\frac{1}{4}$ inch thick, laid flat in the ground with the edge on one side touching the wall, or vertically underneath the edge of the pavement kerb, at a depth of 6 inches below the surface, and placed end to end in contact with each other; each plate to have riveted in it two round wrought-iron vertical rods, each $\frac{3}{8}$ inch in diameter, 12 inches long, and sharply pointed at the upper extremity; the rods to be fixed at 2 feet distance from each other, 1 foot from each end of the plate respectively, and 1 inch from the edge of the plate's inner side; the soil excavated to be compactly refilled over the plates to the original level, so as to allow of the upper 6 inches of the rods to project above the surface of the ground; the exposed portions of the rods to be painted or tarred.

The weight of such a plate made at Devonport was 20$\frac{1}{2}$ lbs., and the cost of it was 5s. 7$\frac{1}{4}$d.; thus the expense of the system, including all labour, may be put at 1s. 6d. per running foot.

[1] VII. G *c*.

IX. B *d*.

The plates can be prepared by any smith, and can be laid by the occupier of the building without the intervention of an electrician; the points can always be seen and be kept sharp; there is practically no exposure of metal; the building is not disfigured; and whenever it is considered desirable to examine the plates, all that is needed is to remove the small layer of superincumbent earth.

Cast-iron taps, with continuous vertical webs of an ornamental pattern surmounted by a fretwork of points, might be appropriately employed at certain kinds of buildings, in lieu of the wrought-iron form above specified.

Such an apparatus as proposed would, according to our views, exercise a beneficial tapping function in any kind of ground that it was placed in, however rocky or dry the site might be.

Wherever the metal should touch the rock, the electrical contact between the plate and the earth would doubtless be less complete than when a certain amount of moisture was present; but in these rocky situations, according to the theory advanced in Part II., there is proportionally less possibility of a charge accumulating to any extent;[1] and if it should do so, the amount of electricity tapped by the apparatus would be correspondingly increased; in fact, the power of the tap would vary with the explosive condition of the ground; and the more the apparatus should be needed, the more it would respond to the call.

That such a contrivance, or indeed any that can be devised, can be relied on absolutely to prevent explosion, seems quite out of the question; but, so far as we are aware, it certainly is not liable, *quâ* the object for which it is intended, to any source of failure.

Although, in its complete form, the proposed electric tap would surround the whole building, still it is considered that this course would only be needed with buildings like powder magazines, which require extra-

---

[1] VII. A *d*.

ordinary precautions, though it might conveniently be applied also to others which only occupied a comparatively small area, such, *e.g.* as monuments.

When applied, then, to buildings in general, the tap would usually merely embrace either the base of any prominently elevated feature needing special protection —*e.g.* the spire or tower of a church, the chimney-stalk of a furnace—or the pavement enclosing that base, and where such feature did not reach the ground on all sides, the length of tap to be used might be approximately regulated by the same rule as that already suggested for the length of the inside girth of the pavement proposed to be laid adjacent to the feature's imaginary base, and the tap would be disposed over the length or lengths of ground lying close to the walls or pavement kerbs nearest to such imaginary base; but the exact arrangement of the tap would of course depend on the circumstances of the building.

It might occasionally be considered sufficient if only the salient angles of the base or imaginary base of an elevated feature were each guarded by a short length of tap.

Where any particularly elevated features were wanting, those most prominent laterally, *e.g.* the corners or projecting portions, might be protected.

In cases where the building is enclosed with pavement as proposed, it is considered that the efficacy of the apparatus in tapping the electricity at the base of the walls would not generally be affected by the short distance the tap would be kept therefrom, necessitated by the width of the pavement; but should the nature of the ground lead to the idea that this might possibly be the case, the pavement would have to be omitted.

The following constructions would, as a rule, be supplied with electric taps, viz. :—

> (1.) Buildings for manufacturing, manipulating, or storing gunpowder or other explosives.

These would be the only kind of buildings that

IX. B *d*.
(it is proposed) should, of necessity, be completely surrounded by the taps.

At the rate of 1s. 6d. per lineal foot, the cost of defending an existing magazine holding 10,000 barrels of powder (500 tons), the girth of the pavement around which is 474 feet, would be £35/11.

The actual cost of defending a similar magazine, holding only 2,000 barrels, on the system now in force, has been found to be £112/2/11.

(2.) Tall chimney-stalks of furnaces.
To be dealt with *quâ* the stalks only.

(3.) Monuments, columns, and works of art.
To be surrounded when of small area.

(4.) Churches and other buildings with prominently elevated features.
To be dealt with *quâ* the elevated features only.

(5.) Large stores, warehouses, and factories, in exposed positions.
These would generally not present any elevated features, and the prominent corners should in such cases be protected for a short distance on each side of the angle, and especially those most exposed to the winds usually accompanying the thunderstorms of the locality.

Ordinary country houses, and country labourers' cottages are not intended, as a rule, to be provided with these taps.

In the case of country houses, however, it would probably, as we have already stated *in re* metals, be a question dependent on circumstances whether such special protection would be desirable or not; and since the expense of the proposed taps would be small, doubtless it might sometimes be worth while to employ them at such houses, and especially in cases where the removal of external elevated metal should cause no inconvenience.

IX. B e.

We have mentioned the manner in which we consider that the ground around the houses of large towns is already tapped by means of their gas and water pipes; and of course houses in the country supplied with either of these services would be to a great extent in a similar condition.

(*e*) *Summary of Proposals for the Defence of Buildings.*

The following is a summary of the arrangements recommended for the defence of buildings of various kinds from the effects of lightning; but, provided that pavement was never employed without the removal of external metal, it would generally depend on the circumstances of the particular building whether all of the proposed measures would be needed, or whether the adoption of some one or more of them would be sufficient.

(1.) *Buildings connected with Gunpowder or Explosives.*

(*a*) Choose a rocky or dry site, remote from the banks of rivers and small sheets of water, and as little exposed as possible. When practicable and convenient, select an underground site for a magazine.

(β) Avoid elevated features in the building, and keep it as low as possible.

(γ) Omit all metal on its external surface, in the body of the walls, and adjacent to the outer walls inside the building.

(δ) Pave the strip of ground immediately around the building.

(ϵ) Lay an electric tap in the ground around the building.

(2.) *Chimney Stalks of Furnaces.*

(*a*) Omit all metal on the external surface, and, so far as practicable, in the body of the walls and inside the shaft.

IX. B e.

  (β) Pave the ground adjacent to the base of the stalk.

  (γ) Convert the metal work of the furnace into an electric tap.

  (δ) Lay an electric tap in the ground adjacent to the base of the stalk.'

   (3.) *Monuments, Columns, and Works of Art.*

  (α) Avoid surmounting with metal in the form of a statue, trophy, or other work of art; and omit all metal on the external surface.

  (β) Pave the ground all around the pedestal of the work.

  (γ) Lay an electric tap in the ground around the work.

(4.) *Churches and other Buildings, with Prominently Elevated Features, not included in* (2) *or* (3.)

  (α) Omit all metals as in (1) at the elevated features, and all external elevated metals at the rest of the building.

  (β) Pave the ground all round the building, or adjacent to the bases of the elevated features.

  (γ) Lay electric taps in the ground adjacent to the bases of the elevated features.

(5.) *Large Stores, Warehouses, and Factories, in Exposed Positions, not included in* (4.)

  (α) Keep the amount of external elevated metal at a minimum.

  (β) Pave the ground all around the building or adjacent to the prominent angles.

  (γ) Lay electric taps in the ground adjacent to the prominent angles.

   (6.) *Country Labourers' Cottages.*

(α) Omit all external elevated metals.

(β) Arrange the grates on the lowest floors as electric taps.

(7.) *Country Houses, not included in* (4) *or* (5), *and not otherwise considered to need special protection.*

(α) Omit external elevated metals so far as convenient.

(β) Arrange the grates on the lowest floors as electric taps.

(8.) *Farm Buildings.*

Omit external elevated metals so far as convenient.

(9.) *Town Houses where gas or water is laid on, not included in* (4) *or* (5.)

Nil.

(*f*) *The Defence of Coal Mines.*

The dangers possibly due to the deep shafts of mines have been already alluded to;[1] and it seems quite possible that the recent terrible accident at the Risca Colliery during a severe thunderstorm may have been caused by an accumulation of thunderbolt explosive conditions at the bottom of the shaft, and by the absence of restraint caused by the sides of the shaft.[2]

It is suggested that coal mines should always be treated, *quâ* lightning, as stores of "fire-damp," and should, in like manner as stores of other explosive substances, such, *e.g.* as gunpowder, be defended by all known means from being ignited by lightning.

Coal mines would seem in fact to be far more in need of such protection than powder magazines; for the former generally contain at all times of the day and night a great number of human beings, whilst the latter seldom contain any, and never when a thunderstorm is known to be in progress.

In dealing with mines, we are as yet without actual experience; for there is apparently no case on record of a

---

[1] VII. B*f.*

[2] It appears from the Report of the Lightning Rod Conference that on the 12th July, 1880, lightning actually entered the workings of Tanfield Moor Colliery.

IX. B *g*.

mine positively known to have been exploded by a thunderbolt; still, every reasonable precaution would probably be worth adoption; and if the theory of the proposed electric taps be sound, much benefit might arise from the use of them at the bottom of deep shafts.

It seems quite possible that the coal at the foot of the shaft may be able to collect electricity with some rapidity; for carbonaceous substances appear to rank after metals in their influence as collectors; and if we once allow that lightning ascends from the earth instead of descending on it from the clouds,[1] there is good reason for presuming that mines may occasionally be the scenes of thunderbolts.

(*g*) *The Defence of Ships.*

We do not propose to make any detailed suggestions regarding the defence of ships, and for the following reasons, viz.:—

> (1.) Iron ships, whether carrying lightning rods or not, are apparently never struck by thunderbolts.[2]
>
> (2.) The form of lightning rod in use in H.M. ships appears to have been (though not without exceptions)[3] successful in tapping any electricity that may have accumulated around them.[4]
>
> (3.) To judge from "Lloyd's List," lightning accidents at sea seem, at the present time, to be rare;[5] nevertheless, it appears that the larger wooden merchant ships seldom carry lightning rods, and the smaller ones never; hence the inference is that, in these days, thunderbolts do not often occur at sea, and certainly not so frequently as formerly.[6]

It is submitted, however, that the principle of electric taps, as proposed for buildings, could readily and ad-

---

[1] VI. B *b*.   [2] VII. B *d*.   [3] III. 69, 70, 99—102.
[4] VIII. B.   [5] III. 125.   [6] I. G 1—4.

IX. B *g;* C *a.*

vantageously be applied to all kinds of ships; for all that they need for this purpose is the application of short points at convenient places inside the hull in connection with the iron or coppered bottom; and this would permit of the dangerous and costly copper bands and tubes, fixed to the masts and shrouds, in the present system of ships' lightning rods,[1] being dispensed with.

(C.) THE DEFENCE OF INDIVIDUALS.

(a) Rules for the Guidance of Individuals.

As regards the defence of individuals from thunderbolts, much good would probably result from the knowledge, and carrying out, of a few simple rules.

It is submitted that one of the forms of Government supervision might, in rural districts, advantageously consist in circulating printed directions on the subject of defence from lightning.

Pending the obtaining of further experience, the following suggestions are now offered for the guidance of individual action during the progress of thunderstorms:—

Inside Houses.[2]

(1.) Whenever not inconvenient, vacate kitchens and all rooms on the lowest floor where there are fireplaces.

(2.) Where this is impracticable or inconvenient, carefully avoid the neighbourhood of such fireplaces.

(3.) In all rooms, keep, as a rule, clear of the fireplaces and of the outer walls; and remain as much as possible in the middle of the room.

(4.) Especially avoid the vicinity of any metals on or near the outer walls, such, *e.g.* as balconies, rain-water pipes, window bars, iron shutters and doors, gas-pipes, water-pipes, tie rods, bell wires,

---

[1] III. 69. IV. 44.   [2] I. F 17.

IX. C *a*.
speaking tubes, safes, cisterns, sinks, large mirrors, gildings, and bedsteads.

(5.) Keep all windows, doors, and other openings closed.

(6.) An underground vault or cellar is usually a secure place.[1]

(7.) Always keep chimney flues, and especially the kitchen flue, fairly clear of soot.[2]

In the Open Air.

(1.) If you be about to walk, ride, or drive, in the country, during the summer or autumn in thunderous weather, do not carry more metal in any form about your person than is absolutely necessary.[3]

(2.) Under the above circumstances, when walking, take an umbrella with you with as little metal on it as practicable, and not a walking stick.[4]

(3.) If overtaken by a thunderstorm get as soon as possible inside any masonry building that may be near,[5] or failing a house of this kind, the nearest house of any sort; but avoid wooden sheds and out-buildings.[6]

(4.) If there be no houses near, do not attempt to obtain shelter anywhere; but choose the ground near you that appears to be naturally the least exposed and the driest, and that is not close to water of any kind, and sit or recline there covered by your umbrella.[7]

(5.) Avoid especially the neighbourhood of trees,[8] hedges, fences, walls,[9] steep faces of rock, and all similar shelters.[10]

(6.) If overtaken when riding, dismount;[1] and, if practicable, leave or secure the horse standing on a site similar to that mentioned in (4), and sit or recline in another similar position yourself, some little distance away.

(7.) If overtaken when driving,[2] stop, and if in a covered carriage, dismount from the outside, and either get inside or take up a position, as in (4), at some distance from it. If the conveyance be uncovered, all should dismount and dispose themselves as in (4), at some distance off.

Some authorities have recommended the portions of ground near, but not close to, shelters, such as trees, as being safe positions, in comparison to other places, on the principle that the shelter itself would as a rule be struck by any thunderbolt that might occur thereabouts, and thus the ground near the shelter, but not immediately contiguous, would probably be rendered comparatively safe.[3]

We must, however, remember that this neighbourhood —and especially if a human being were in it—would, if a thunderbolt struck the shelter, generally be the scene of a return stroke; and therefore, though life might be secure there, still it would probably be at the cost of a more or less severe shock; moreover, it would be impossible to estimate accurately the distance from the tree or shelter at which immunity from a thunderbolt striking the tree might end, and danger from other thunderbolts might begin.

(b) *Agricultural Labourers.*

The persons who appear to be most exposed to the effects of thunderbolts are agricultural labourers.[4]

They are frequently working in the fields when a thunderstorm comes on, and they naturally seek refuge at

---

[1] VII. B *e;* C 8, η.   [2] VII. B *e;* C 8, η.   [3] I. F 11, 12.
[4] I. F 9; G 8, 9, 19—21, 36.   VII. C 8, δ, ι, ν.

IX. C *b*.

the nearest shelter; but this shelter is generally a tree or slight wooden shed, and, so far from affording them protection from lightning, is constantly the cause of their deaths by it.[1]

The remedy that we propose is the substitution of masonry for wood as the material for the construction of field sheds, and the use of these sheds in greater numbers.

The labourers should be careful to leave their tools where they have been working.

[1] I. F 10.   VII. C 8, λ, μ.

INDEX.

ABEL, Professor, 34, 71
   Acceleration, 3, 148
Accidental dielectrics, 201, 220, 260
Adair, Mr., 93
Afghanistan, 44
Africa, South, 79, 113, 128
Agency, 145
Agram, 25
Aide - Mémoire to the Military Sciences, 12, 24, 25, 51, 73, 128
Air, 6, 7, 9, 10, 14, 28, 30, 61, 173, 189, 199, 204, 222, 290
Airth, Alexander, 124
Almshouses, 142
Amber, 47
America, North, 43, 45, 49, 58, 64, 73, 87, 89, 90, 93, 102, 103, 104, 107, 109, 110, 112, 115, 123
America, South, 45, 102, 108, 111
Ampère, M., 18
Analysis of incidents, 209, 258
Anderson, Joseph, 126
Anderson, Mr. R., 10, 13, 15, 16, 24, 29, 35, 36, 43, 44, 46, 48, 49, 50, 51, 52, 57, 58, 64, 68, 69, 70, 73, 74, 75, 76, 82, 83, 84, 99, 102, 106, 116, 117, 118, 119, 120, 121, 122, 123
Angularities, 152
Animals, 7, 9, 10, 31, 32, 39, 173, 192, 194, 218
Apennines, 45
Arago, Professor, 12, 13, 14, 15, 21, 23, 24, 25, 26, 27, 28, 29, 30, 31, 32, 36, 37, 38, 39, 43, 44, 45, 50, 52, 59, 65, 72, 73, 77, 86, 87, 88, 89, 90, 91, 92, 93, 94, 95, 99, 103, 104, 105, 107, 162, 223
Architects, 243, 273
Architectural arrangements, 257, 272
Architecture, works on, 74
Areas, large, 267
Articles of dress, 171, 172
Artillery, 38
Ashes, 7, 67, 68, 172
Asphalte, 71, 183, 184, 272, 276
Asphyxia, 33
Associated dielectrics, 220
Atlantic Ocean, 22, 86, 103, 108
Atmosphere, 7, 8, 11, 14, 15, 20, 46, 63, 80, 83, 154, 220, 231
Attraction, 1, 5, 6, 13, 35, 47, 80, 83, 146, 166, 199, 225, 226
Augustus, Emperor, 36, 37
Auroræ, 16, 17, 18, 19, 20, 21, 156, 159, 175, 232, 269
Austria, 43, 86, 92, 104
Aylesford, Lord, 87

BALCOMBE (a boy), 131
   Balconies, 171, 272, 289
Balfour Stewart, Professor, 16, 19, 20
Balloons, 12
Balls, 272
Balustrades, 171
Barns, 45, 73, 211
Batteries, coast, 71

Beccaria, M., 50
Becquerel, M., 10, 52
Beds, 43, 217
Bedsteads, 172, 173, 290
Beer, 12, 228
Belfries, 200, 205, 274
Bellion, Mr., 90
Bells, 39, 43, 171, 205, 273, 275
Bell wires, 30, 289, 292
Béranger, M., 88
Berlin, 45
Bernouilli, John, 47
Bevis, Dr., 48
Bismuth, 9
Bituminous substances, 7, 173
Blomfield, Mr., 111
Boats, 173
Boilers, 171
Bone, 173
Boyle, Robert, 47
Bracini, M., 25
Brass, 56, 171, 213
Bricks, 82, 173, 184, 193, 194, 272, 276
British Association, 2, 265
Bronze, 171, 213
Brush discharge, 6, 64
Brussels, 54, 60, 65
Buchanan, Mr., 82
Bucharest, 25
Buffon, M., 48
Buildings, 30, 31, 35, 36, 70, 71, 72, 74, 77, 79, 80, 83, 84, 171, 172, 173, 174, 184, 199, 203, 210, 246, 262, 265, 271, 290
Buildings set on fire, 211
Bushes, 172
Butter, Mr. E., 123
Buys, Mr., 113

CAIRO, 45
Callaud, M., 65, 69, 73
Campaniles, 273
Canvas, 173
Caoutchouc, 10
Capacity, 2, 3, 145, 149, 150, 151, 153, 167, 185, 194, 199, 200
Carbonaceous substances, 172, 288
Carriages, 173, 291
Carts, 173, 291
Cast-iron electric taps, 282

Castle, Mr., 112
Castles, 275
Cattle, 173
Causes of lightning rod failures, 259
Caves, 36, 203
Cavendish, Mr., 6, 49, 59, 92
Cellars, 290
Cement, 173, 276
Chalk, 7, 8, 9, 173
Chambers's Dictionary, 26
Chappe, Professor, 13, 27
Charcoal, 7, 8, 9, 10, 65, 69, 172, 252
Charcoal trenches, 187
Charge, 176
Chatham notes, 5
Chemical action, 2
Chili, 24, 25, 44, 160
Chimneys, 38, 44, 70, 71, 79, 206, 215, 278, 279, 290
Chimney-pots, 171, 200, 207, 272
Chimney-stalks, and shafts, 141, 142, 172, 200, 207, 208, 274, 278, 280, 284, 285
Chittenden, Mr., 131
Churches, 35, 50, 66, 84, 137, 138, 139, 140, 141, 142, 143, 144, 210, 272, 274, 278, 284, 286
Cisterns, 65, 67, 290
Clarke, Samuel, 126
Clay, 167, 173
Clerk, Maxwell, Professor, 83
Cliffs, 201
Clocks, 171, 205, 273, 275
Clothes, 171, 172, 197, 218
Clouds, 11, 12, 14, 15, 16, 27, 28, 29, 31, 46, 63, 80, 81, 154, 156, 157, 173, 176, 221, 223, 224, 225
Cloud explosions, 175, 223, 225, 228
Coal mines, 46, 201, 287
Coast, sea, 180
Coke, 7, 65, 66, 172, 252
Collecting plates, 147, 152, 156, 160, 191
Collection, 147
Collective substances, 179, 181
Collectors, 146, 150, 171, 172, 173, 288
Columns, 138, 171, 200, 258, 273, 274, 284, 286

INDEX. 295

Comazants, 22
Combustion, 11
Concrete, 71, 173, 272, 273, 276
Condensation, 5
Condenser, 5, 147, 150, 151, 152, 155
Condenser, terrestrial, 154, 170, 174, 203
Condensing plate, 147, 152, 191
Conduction, 48, 80, 147
Conductivity, 6, 9
Conductors, 4, 5, 9, 10, 146, 171, 172, 173
Conservatories, 173
Constructions, 188, 203, 211, 259, 270
Contact, 189, 190
Copper, 6, 7, 8, 9, 10, 52, 53, 54, 55, 56, 57, 58, 59, 63, 64, 68, 83, 171, 213, 249, 250, 255, 272, 274
Corn, 43, 173
Cost, 51, 70, 74, 249, 250, 251, 252, 253, 265, 281, 284
Cottages, 207, 208, 211, 274, 275, 278, 279, 284, 286
Cotton, 8, 10, 173
Countries, defence of, 267
Country, 45
Country houses, 208, 278, 279, 284, 287
Cows, 173
Cramps, 171, 273
Crosses, 171, 272, 274
Culley, Mr., 16, 68
Cupolas, 274
Current, 3, 18, 19, 20, 33, 49, 58, 146, 147, 149, 165
Cuvier, Professor, 24

D'ABBADIE, 13, 29
D'Alibard, 12, 48, 49, 241
Dam of reservoir, 152, 153
Dangers to interiors, 204, 206
Davy, Sir Humphrey, 10
Deaths, 31, 32, 33, 40, 41, 42, 43, 44, 292
Decomposition, 11, 228
De Fonvielle, M., 75
De la Rive, Professor, 20
De la Rue, Dr. Warren, 58

De L'Isle, M., 13
Della Torre, M., 25
Denmark, 120
Density of atmosphere, 231
De Romas, M., 50, 241
Deschanel, Professor, 3
Deviations of explosions, 261
Dielectrics, 5, 147, 152, 167, 191, 204, 220
Dillwyn, Mr., 14
Distance, 148
Districts, defence of, 267
Districts, rural, 289
Discharge, electric, 6, 11, 59, 61, 83, 146, 148, 150, 151, 152, 153, 174, 175, 176, 177, 178
Diversities of opinion, 243
Doors, 173, 205, 289, 290
Dogs, 173
Donkeys, 173
Drainage, 84
Drain pipes, 70, 171
Driving, 198, 290

EARTH, 7, 9, 10, 14, 15, 16, 17, 19, 27, 28, 63, 80, 154, 156, 157, 177
Earth, "bad," 173, 259
Earth, dry, 66, 67, 68, 69, 173, 182, 189
Earth, "good," 173, 182
Earth, moist, 65, 66, 67, 68, 69, 81, 162, 173, 182
Earth connections, 56, 65, 66, 67, 68, 69, 70, 71, 75, 79
Earthquakes, 17, 23, 24, 25, 26, 44, 160, 269
Earth currents, 16, 17, 20, 159, 269
Earth's crust, 186
Earthenware, 173, 273
East India Company, 77
Eclair, l', 176
Electricity, 1, 2, 3, 4, 5, 6, 16, 145, 146, 147, 148, 149, 150, 151, 152, 153, 155
Electricity, atmospheric, 10, 20, 223
Electricity, negative, 4, 11, 15, 63, 145, 166
Electricity, positive, 4, 11, 63, 145, 166

Electricity, terrestrial, 10, 14, 21, 82, 156, 157, 179
Electricity, thermal, 2, 11, 18, 19
Electricity, voltaic, 1, 11
Electric connection, 188, 189
Electric fluid, 50, 62, 81
Electric machine, 47
Electric sparks, 6, 28, 47, 48, 166, 234, 241
Electric taps, 278, 279, 280, 281, 282, 283, 284, 285, 286, 287
Electricians, 69, 74
Electrified bodies, 1, 2, 4, 145
Electrolytic action, 65, 66, 190
Electro-motive force, 3, 146, 149
Elevated features, 274, 278, 283, 284, 285, 286
Elevated positions, 187, 274, 286
Elevation, 72, 185, 194, 200, 276
Encyclopædia Britannica, 23, 24
Encyclopædia Metropolitana, 44
Energy, 1
Engines (railway), 172
England, 39, 40, 41, 42, 44, 55, 59, 69, 74, 76, 87, 90, 91, 92, 93, 94, 99, 100, 102, 103, 104, 106, 107, 108, 109, 110, 111, 113, 114, 115, 116, 117, 118, 119, 120, 121, 122, 123, 124, 126, 127, 128, 129, 130, 131, 132, 133, 137, 138, 139, 140, 141, 142, 143, 144, 173, 222, 240, 242, 243, 267, 269
Eruptions, volcanic, 17, 24, 25, 26, 156, 160, 269
Etna, Mount, 26
Evaporation, 11
Everett, Professor, 10
Explosion, 30, 146, 147, 150, 153, 165, 176, 177, 202, 208
Explosive action, 190, 191
Explosive conditions, 271, 276
Exposed positions, 187, 274, 286

FACILITATION, 147
Facilitators, 171, 172, 173
Facilitators, great, 171, 172
Facilitators, slight, 173, 191
Factories, 274, 275, 284, 286
Factories, gunpowder, 271, 274, 283, 285

Faraday, Professor, 61, 78
Farm-houses, 275, 279, 287
Fat, 8
Feathers, 7, 8, 9
Felt, 173
Fences, 171, 173, 190
Fields, 182
Finials, 171, 172
Fires, 38, 44
Fire balls, 229
Fire damp, 287
Fire-places, 38, 71, 279
Flagging, 277
Flagstaffs, 71, 162, 173, 194, 216
Flame, 7, 9, 63, 172, 207
Flammarion, M., 13, 20, 21
Floors, 38, 171, 173, 273
Fog, 44, 173
Foliage, 173
Force, 1, 3, 148
Force, expansive, 30, 31, 191
Formulæ, electric, 148
Foudre, la, 176
France, 24, 26, 38, 39, 43, 45, 50, 57, 58, 59, 64, 74, 76, 86, 87, 88, 90, 91, 92, 94, 98, 123, 126, 176, 222, 242
Francisque Michel, M., 56, 62, 68, 76.
Franklin, Benjamin, 5, 12, 30, 37, 38, 48, 49, 56, 59, 60, 63, 69, 70, 74, 78, 92, 102, 238, 240, 241, 242, 248
Fraser, Lieut. T., 26
Friction, 11, 12
Fulgurites, 30
Fur, 7, 9, 173
Furnaces, 172, 208, 274, 280, 284, 285
Fusion, 30, 192, 260

GALTON, Capt. Douglas, 71
Galvanometers, 75, 255
Galvani, Professor, 31
Galvanic action, 255
Ganot, Professor, 1, 5, 6, 9, 10, 13, 16, 17, 18, 21, 22, 23, 27, 32, 37, 56, 57
Gaps, air, 50, 166, 241
Gas, 7, 8, 9, 10
Gasholders, 206

INDEX. 297

Gas-pipes, 66, 67, 73, 171, 205, 235, 251, 253, 270, 273, 276, 285, 287, 289
Gas works, 142
Gates, 171
Gavarret, Professor, 60
Gay-Lussac, Professor, 51, 173
Geography, 39, 42, 43, 44, 45
Geological formation, 185, 186
Germany, 59, 76, 90, 105, 116, 119, 120
Gilbert, Dr., 17, 47
Gildings, 38, 62, 213, 290
Girders, iron, 171
Glass, 6, 7, 8, 9, 10, 30, 38, 47, 60, 173, 200, 273
Gold, 9, 10, 171
Gordon, Mr., 1, 5, 10, 17, 19, 64
Government, 267, 268, 269, 289
Graphic newspaper, 23, 46
Graphite, 7, 9, 10, 172
Grass, 173, 233
Grates, 70, 171, 207, 278, 286, 287
Gratings, 273
Graves, Mr., 28, 85, 115
Gravity, force of, 225, 226
Gray, Mr. Stephen, 48
Griswold, Mr., 23
Ground, 11, 14, 15, 179, 185, 187, 190, 281
Guillemin, M., 82
Gunpowder, 72, 91, 105, 117, 128
Gutta-percha, 6, 7, 10
Gutters, eaves, 73, 171, 172, 173, 200, 272, 273

HAIL, 12, 13, 84, 172, 221
Hair, 7, 8, 9, 173
Halley, Professor, 19
Hamilton, Sir Wm., 25
Halls, 275
Hare, Professor, 65
Harris, Sir Wm. Snow, 1, 6, 7, 13, 15, 22, 35, 39, 51, 52, 53, 55, 60, 72, 77, 78, 79, 86, 88, 89, 92, 94, 95, 97, 98, 99, 100, 101, 102, 103, 104, 105, 107, 108, 109, 110, 162, 244, 249
Hartley, Mr. Thomas, 92

Hawksbee, Mr. 47
Hay, 173
Heat, 2, 16, 23, 191, 192
Heathcote, Mr. R. B., 46
Hedges, 173, 290
Hemp, 173
Henley, Mr., 6, 37
Herschel, Sir John, 11, 12, 13, 29, 30, 50
Hills, 185
Historical notes, 47, 238
Holtz, Professor, 36
Horses, 173
Houses, 173
Howorth, Mr., 24
Human bodies, 173, 192, 193, 194, 196, 202, 216, 217, 218, 229
Humboldt, Herr Von, 18
Huts, 173
Hydraulic simile, 152

ICE, 7, 8, 9, 10, 13, 84, 172, 183
Iceland, 44
Illustrated London News, 111
Incidents, lightning, 86
India, 44, 103, 109, 113
Indian Ocean, 101
India-rubber, 6, 8
Individuals, defence of, 289
Induction, 1, 4, 5, 32, 33, 79, 147, 155
Inductive capacity, 5, 150
Influence, 147, 170, 171, 172, 173
Insulation, 4, 14, 48, 147, 270
Insulators, 4, 8, 9, 73, 74, 81, 146, 150, 151, 173, 189, 244, 257
Interiors of buildings, 203, 204, 212, 273, 289
Ireland, 41, 109, 115, 128, 129
Iron, 6, 9, 10, 17, 35, 45, 49, 54, 55, 56, 57, 58, 59, 64, 65, 68, 70, 71, 82, 171, 172, 213, 214, 272, 273, 279, 288
Iron-pointed plates, 281
Iron ships, 288
Iron spikes, 278, 279
Italy, 45, 50, 87, 89, 91, 93, 95, 99, 104, 105, 110, 119, 122
Ivory, 173

JAMAICA, 44, 123
  James, William, 124
Jenkin, Professor F., 2, 3, 4, 5, 6, 7, 15, 16, 19, 28, 31, 64
Joists, iron, 171

KAEMTZ, Professor, 11, 12, 13, 27, 43, 81
Kæmpfer, M., 36
Kentish paper, 107, 129, 130, 131, 132
Kitchens, 207, 289
Kites, 50, 240
Kleist, Herr, 48
Kriel, Professor, 24

LABORATORIES, gunpowder, 271, 273, 285
Labourers, 40, 41, 274, 275, 291
Lakes, 182, 187
Lambert, M., 13
Latimer Clark, Mr., 8, 29, 55, 71, 85, 115
Law, electric, 149, 165
Lead, 6, 9, 10, 70, 71, 73, 172, 213, 272, 273
Leaks, 4, 146, 148, 151, 153, 176, 194, 270
Leaks, atmospheric porous, 231, 232
Leaks, terrestrial, 175, 231, 232, 233, 234, 235, 236, 237
Leather, 7, 8, 9, 173, 197
Le Gentil, M., 13
Lenz, Herr, 10
Le Roy, M., 116
Leyden jar, 5, 48
Lieberkühn, Dr., 48
Lighthouses, 60, 211
Lightning, action of, 30, 86
Lightning, ascending, 27, 28, 161, 264
Lightning, ball (globular), 26, 29, 229
Lightning conductors, 36, 63, 80, 81, 82, 83, 85, 242
Lightning, definition of, 26, 28
Lightning, descending, 27, 28, 160, 169, 242

Lightning discharge, 15, 25, 26, 27, 28, 29, 46, 47, 48, 174, 176
Lightning engineering, 47
Lightning flashes, 11, 12, 26, 29, 49, 50
Lightning, heat, 28, 156, 175, 232
Lightning incidents, 86, 209
Lightning, mechanical force of, 163, 164, 168, 193, 219
Lightning protectors, 34, 230
Lightning, sheet, 26, 28, 156, 175, 232
Lightning, zigzag, 26, 29, 191
Lightning rods, 48, 49, 50, 53, 73, 171, 202, 238, 288
Lightning rods, action of, 239, 245, 261
Lightning rods, application of, 70, 71, 72, 73, 74
Lightning Rod Conference, 85, 265, 287
Lightning rods, cost of, 249, 250, 251, 252, 253, 265, 284
Lightning rod details, 51, 52, 53, 54, 55, 56, 57, 58, 76, 137, 244, 254
Lightning rods, disadvantages of, 246, 280
Lightning rods, disfiguring tendencies of, 257
Lightning rods, fused ends of, 260
Lightning rods, history of, 47, 238
Lightning rod incidents, 258, 261, 262, 263
Lightning rod inspections, 75, 76, 255, 256, 257
Lightning rods, instances of, 137
Lightning rods, mechanical injuries to, 260
Lightning rod points, 54, 59, 60, 61, 62, 63, 64, 75, 79, 238, 250, 251, 253, 256, 257, 264, 265
Lightning rods, protective power of, 51, 77, 78, 79, 80, 81, 82, 83, 84, 85
Lightning rods, roots of, 238, 245, 246, 247, 251, 252, 256, 264, 265
Lightning rods, stalks of, 238,

244, 247, 248, 251, 254, 255, 264, 265
Lightning rods, sources of failure of, 264
Lightning rods struck, 50, 51, 77, 78, 163, 164, 169, 193, 243, 258
Lightning strokes, 32, 78
Lightning strokes, accurately defined, 219, 230
Lightning strokes, divided, 29
Lightning strokes, horizontal, 219
Lightning strokes, repeated, 219
Lightning strokes, simultaneous, 208, 218
Lime, 7, 9, 10, 173
Limestone, 14, 173
Linen, 8, 10, 173
Lloyd's List, 288
Local dielectrics, 189, 192, 193, 198, 207, 208, 220, 260
Local plates, 188, 191, 192, 194, 196, 219, 260
London, 42, 45, 50, 95, 97, 101, 105, 106, 108, 116, 120, 121, 122, 124, 125, 126, 131, 141
Ludolf, Herr, 48
Lunn, Mr., 8, 27

MACTAGGART, Mr., 82
Maffei, Professor, 27
Magazines, gunpowder, 51, 55, 69, 71, 72, 82, 139, 211, 271, 274, 276, 283, 284, 285
Magnetic storms, 16, 17, 18, 19, 20, 24, 158, 269
Magnetism, 10, 17, 30, 158
Magnetism, the earth's, 16, 17, 18, 19, 21, 157
Mahon, Lord, 31, 50, 162
Majendie, Major (report of), 117
Malaise, 24, 228
Malta, 127
Manilla, 25, 26
Mann, Dr., 13, 14, 29, 32, 35, 37, 43, 53, 54, 60, 61, 65, 66, 73, 75, 79, 113, 114
Marble, 7, 8, 9
Market buildings, 138
Masonry, 81, 212, 277, 290, 292
Mass, 2, 3, 148

Masts, 21, 39, 53, 72, 81, 82, 144, 173, 202, 216
Matthiesen, Professor, 7
Matterhorn, 46
Matting, 172
Mattresses, 38
Maxwell, Mr. Hugh, 37
Meat, 12
Mediterranean, 45, 103, 110
Melloni, Professor, 95
Melsen, Professor, 54, 57, 60
Mercury, 9, 172
Metals, 7, 8, 9, 10, 30, 31, 35, 36, 58, 52, 72, 73, 82, 84, 171, 172, 174, 185, 190, 192, 194, 197, 198, 199, 200, 205, 206, 213, 236, 237, 246, 247, 271, 272, 273, 285, 286, 287, 289
Metal points, 232, 233, 234, 236, 237
Mica, 6
Michælis, M., 107
Milk, 12
Mines, 34, 44, 45, 201, 230, 287
Mirrors, 38, 172, 273, 290
Mist, 46, 173
Moisture, 8, 48, 65, 69, 76, 84, 181, 190, 224, 232, 251, 253
Money (in purses), 171
Monuments, 274, 277, 284, 286
Mortar, 173
Mould, vegetable, 67
Mountains, 44, 45, 46, 185
Mountain summits, 46, 168, 183
Municipal authorities, 269

NAILS, 31, 171, 272
Nelson, Colonel, 12, 73, 128
Newall, Mr., 57, 58, 64
Nickel, 17
Night, 222
Nitric acid, 12
Nitrogen gas, 22
Nollet, Abbé, 31, 49, 77
Non-conductors, 9, 10, 173
Nouel, M., 28

OCCUPIERS of houses, 273, 275
Odour, sulphurous, 28.
Ohm, Professor, 3, 10, 58

Oil, 173
Oil tanks, 73, 172, 206
Openings in buildings, 205
Opposition to lightning rods, 242
Ores, metallic, 7
Organs, 171, 273
Ornaments, 171
Owners of houses, 273

PALMIERI, Professor, 26
Paper, 7, 8, 9, 10, 173
Paratonnerres, 73, 81, 242
Paralysis, 32
Parchment, 7, 9,
Paris, 22, 45, 60, 75, 89, 90, 99, 100, 116
Parliament, Houses of, 57, 76
Patterson, Mr., 59, 65
Pavements, 183, 184, 187, 235, 277, 278, 285, 286
Paving stones, 173
Pekin, 45
Peltier, Professor, 16, 23
Persons, 31, 32, 36, 37, 38, 39, 45, 196, 216, 217, 218, 289
Peru, 18, 44, 160
Petit, M., 29
Philippine Isles, 44
Pickard, Mr., 47
Pigs, 173
Pinnacles, 274
Pipes, hot-water, 171
Pitch, 6, 38, 173
Plate, terrestrial, 167, 169
Plates, iron pointed, 281
Platina, 9, 59, 60, 63, 64, 171, 250
Plaster, 173
Pliny, 25, 36
Plumbago, 7, 9
Points, 59, 60, 61, 62, 63, 64, 152, 232, 233, 234, 236, 237, 278, 281, 289
Ponds, 172, 181
Pools, 187
Porcelain, 7, 8, 9, 173
Portlock, Col. J., 24, 25
Portugal, 25, 109
Potential, 2, 3, 4, 5, 15, 16, 28, 34, 64, 66, 80, 83, 145, 149, 150, 151, 153, 176, 177, 179, 182, 185, 194, 199, 241
Pouillet, Professor, 10, 11, 51, 52, 63, 68, 82
Power, 145
Practical measures, 266
Preece, Mr. G. E., 35
Preece, Mr. W. H., 1, 27, 29, 33, 34, 39, 54, 55, 58, 59, 61, 66, 70, 71, 75, 79, 80, 84, 115
Press, the, 269
Preservatives, 36
Priestley, Dr., 22, 59
Private houses, 139, 141, 143, 210
Prussia, 43, 144
Public buildings, 210
Pumps, 172

QUANTITY, 23, 145, 148, 149, 151, 153, 176
Quartz, 30

RAILINGS, 171
Railways, 138, 172, 173, 185, 187, 218, 236
Rain, 12, 28, 80, 168, 172, 221, 224
Rainwater-pipes, 67, 70, 73, 171, 172, 272, 273, 289
R. E. Aide-Mémoire, 9, 56, 81
R. E. Journal, 127
R. E. Professional papers, 26
Record of thunderbolt incidents, 267
Registrar-General of England, 39, 40, 41, 42
Registrar-General of Ireland, 41
Reily, Mr., 129
Rending force, 193
Repulsion, 1, 5, 146, 225
Reservoir of water, 152, 153
Resin, 7, 8, 9, 10, 30, 47
Resistance, 3, 6, 10, 28, 29, 147, 149
Restrainers, 173, 191
Restraint, 145, 147, 150, 151, 190, 199, 200, 202
Return strokes, 31, 50, 146, 148, 151, 175, 227, 228, 229, 230
Richmann, Professor, 50, 92

INDEX. 301

Riding, 198, 290
Ridges, 171, 172
Riess, Professor, 62
Risca colliery, 46, 287
Rivers, 15, 43, 46, 172, 181, 186, 253, 276
Roberts, Mr., 124
Rocks, 35, 45, 167, 173, 182, 183, 187, 276, 282
Rocky ground, 276, 285
Rods, tie, 273
Roofs, 171, 172, 173, 200, 216, 272
Roget, Professor, 8
Russia, 25, 44

SABINE, Colonel, 16, 18
Sabrina, isle of, 24
Safes, iron, 273, 290
Salt, 8
Sand, 15, 30, 66, 67, 173, 187
Sap, 30, 162
Sash weights, 171
Schelthorn accident, 198
Schleswig-Holstein, 45
Scotland, 88, 98, 103, 125
Sea, 22, 67, 70, 172, 181, 186
Sealskin, 37
Seamen, 39, 53
Secchi, Father, 122
Seneca, 76
Shape of ground, 185
Shaw, Dr., 33
Sheds, 173, 211, 290, 292
Sheep, 173
Shingle, 15, 66, 67, 173
Ships, 51, 52, 53, 70, 79, 172, 173, 181, 194, 202, 212, 246, 248, 263, 288, 289
Ships, H.M., 39, 45, 51, 52, 53, 144, 249, 288
Ships, merchant, 22, 53, 288
Shock, 3, 32, 175, 191
Shutters, 171, 173, 205, 289
Siemens, Dr. Werner, 34
Silk, 7, 8, 9, 10, 38, 173
Silver, 7, 9, 10, 61, 64, 171, 250
Silver, German, 9
Skylights, 173
Slate, 31, 173, 272
Smoke, 7, 9, 172, 207, 231, 235
Smyrna, 25, 44

Snow, 7, 9, 10, 46, 172, 183, 221
Society of Arts Journal, 113, 114
Society of Telegraph Engineers' Journal, 10, 11, 23, 28, 29, 34, 35. 55, 71, 85, 114, 115, 123, 128
Soil, 174, 187
Soldiers, 218
Soot, 70, 79, 172, 207, 290
Spain, 25, 115
Spence, Dr., 48
Speaking tubes, 171
Spindles, 171, 272, 275
Spires, 22, 31, 54, 81, 137, 138, 139, 140, 141, 142, 143, 144, 173, 193, 200, 205, 272, 273, 274, 278
Springs, 66, 172
Staircases, 171, 173, 273
Statistics, 39, 40, 41, 42, 43, 44
Standard newspaper, 17, 25, 26, 46, 90, 92, 104, 116, 121, 132
Statues, 171, 274
Steam, 7, 9
Steel, 17
Steeples, 22, 50
St. Elmo's fires, 21, 22, 175, 234
St. Helena, 44
St. Petersburg, 45, 50, 92
Stobart, Mr. H. S., 121
Stone, 7, 8, 9, 30, 31, 47, 173, 193, 194, 272, 273, 276
Strain, 5
Stores, 274, 275, 284, 286
Straw, 8, 173
Streams, 181, 187
Subterranean thunder, 14, 25
Suetonius, 36
Sun, the, 16, 19, 20
Sun spots, 19
Surface, 150, 151
Surface of the earth, 179, 187, 194, 201, 221
Surfaces, moist, 167, 168
Surfaces, rocky, 167, 168, 184
Survey, electro-geological, 268
Sweden, 44
Switzerland, 43, 93, 107
Symmer, Professor, 50

302 INDEX.

Symons, Mr., 39, 114
Symons' Meteorological Magazine, 23, 116

TABLE-LANDS, 185
    Tanfield Moor colliery, 287
Tanks, 67, 68, 181, 253
Tar, 173, 276
Telegraphs, 16, 20, 33, 34, 68, 171, 173, 194, 216, 230, 236, 253
Telegraphic Journal, 10, 11, 17, 19, 24, 28, 56, 59, 62, 68, 76, 84, 98, 99, 105, 106, 110, 111, 112, 115, 124
Temperature of the air, 222
Tents, 37, 211
Terra cotta, 173, 272, 273
Testings, 75, 76, 255, 256, 257
Theory, 265
Thickness, 150, 151
Thomson, Sir William, 10, 15, 17, 19, 20, 265
Thorns, 62, 233
Thunder, 22, 25, 28, 44, 46, 47
Thunderbolts, 26, 27, 28, 36, 42, 85, 156, 175, 176, 177, 187, 221, 230, 275, 288, 291
Thunderclouds, 13, 14, 15, 23, 27, 32, 35, 60, 60, 63, 70, 165, 180, 183, 189, 233, 234
Thunderstorms, 10, 12, 13, 14, 15, 21, 22, 24, 31, 33, 34, 37, 38, 44, 45, 46, 154, 159, 226, 290, 291
Tie rods, 289
Tiles, 173, 272
Tillard, Captain, 24
Time, 2, 3, 148
Times newspaper, 79, 124, 125, 126, 127, 129
Tin, 6, 9
Toaldo, Professor, 72
Tomlinson, Mr., 118
Tools, 292
Towers, 137, 139, 141, 142, 143, 144, 200, 205, 273, 274, 278
Towns, 45, 184, 210, 218, 235, 269, 270, 276
Town-halls, 54, 60, 65, 137
Town houses, 276, 287

Trees, 30, 35, 37, 43, 70, 81, 173, 193, 194, 215, 276, 290, 291
Truenfeldt, Mr. Von, 29
Turrets, 273, 274
Tylney, Lord, 89
Tyndall, Professor, 6, 8, 21, 22, 31, 32, 47, 48, 49, 50, 51, 62, 63

UNDERGROUND sites, 285
    Underground vaults, 36, 290
Umbrellas, 290
Uplifting force, 191, 192, 193

VAILLANT, Maréchal, 52
    Valetta, Signor, 25
Valleys, 186, 187
Valves, terrestrial, 235
Vanes, 171, 272, 274, 275
Vapour, 7, 9, 22, 23, 173
Varley, Mr. S. A., 34
Vaults, 36, 290
Vegetable bodies, 7, 8, 9, 10, 173
Vegetation, 11, 173, 182
Velocity, 3, 148
Ventilators, 171
Vesuvius, Mount, 24, 25, 26
Vines, 222
Viollet-le-Duc, M., 81
Volcanic eruptions, 17, 24, 25, 26, 156, 160, 269
Volta, Professor, 11, 32, 38, 239
Von Guericke, Otto, 47
Von Yelin, Herr, 116

WALES, 40, 41, 42
    Walker, Mr. C. V., 17
Walking, 198, 290
Walking sticks, 173, 290
Wallace, Mr., 125
Walls, 38, 72, 203, 206, 290
Ward, Dr., 47
Warehouses, 274, 275, 284, 286
War Office instructions, 6, 15, 30, 31, 35, 51, 55, 56, 62, 66, 67, 68, 71, 72, 75, 80, 81, 123
Watches, 171
Water, 6, 7, 9, 10, 11, 14, 15, 35, 67, 68, 69, 83, 180, 251, 276

Water, head of, 3, 153
Water, rain, 8, 172
Water, sea, 8, 172
Water, spring, 8, 172
Watersheds, 185
Waterspouts, 22, 23, 206, 290
Watson, Dr., 48, 49, 50, 52, 92
Weather, 10, 11, 222
Weathercocks, 171, 200, 272, 274
Weissenborn, M., 29
Wells, 66, 68, 69, 172, 187, 253
Wellington Weekly Gazette, 121, 126, 130
Wellsted, Mr., 131
West, Mr., 102
West Indies, 89, 95
Western Morning News, 106, 108, 120, 121, 129, 130, 131
Western Weekly News, 110, 115, 119, 120, 131, 132, 133
Wheatstone, Sir Chas., 6
Wheel tires, 172

Whirlwinds, 23
Whymper, Mr. E., 46
Whyte, Mrs., 125
Wilson, Mr., 49, 77
Wind, 13, 44, 84, 225, 284
Windows, 171, 173, 205, 273, 290
Winter, 222
Winthrop, Dr., 37
Wires, 30, 192
Wire guards, 273
Wood, 7, 8, 9, 10, 30, 82, 173, 194, 206, 272, 288, 290
Wooden ships, 288
Woodwork, 192, 193, 216
Woodman, Mr. N., 124
Wool, 8, 173
Work, 3, 148, 176

ZINC, 6, 9, 171
Zollner, Dr., 24
Zurich, 81

THE END.

www.ingramcontent.com/pod-product-compliance
Lightning Source LLC
Chambersburg PA
CBHW022023240426

43667CB00042B/1068